AF583706

HOR RID US

JOURNEY OF A TRICERATOPS

First published in 2022 by Museums Victoria Publishing
11 Nicholson Street
Carlton, Victoria 3053, Australia
publications@museum.vic.gov.au
www.museumsvictoria.com.au

Text © Linden Ashcroft, Tony Birch, Terry Ciotka, Maree Clarke, Philip J. Currie, Rebecca Dart, Shannon Faulkhead, Erich M. G. Fitzgerald, Tim Flannery, Dermot Henry, Blair Holtner, Melissa Kay, Nancy Ladas, Benjamin Law, John A. Long, Mitch Mahoney, Susannah C. R. Maidment, Shannon Martinez, Laura Jean McKay, Craig Pfister, Mark Radburn, Hazel Richards, Thomas H. Rich, Shaun Tan, Elaine Vallis, Ry Williams, Tim Ziegler

The moral right of the authors has been asserted.

All rights reserved. No part of this publication may be reproduced, stored in a retrieval system or transmitted in any form by any means, electronic, mechanical, photocopying, recording or otherwise, without the prior written permission of the publishers and copyright holders.

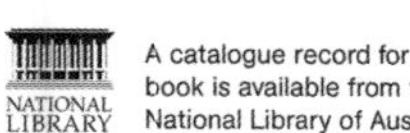

A catalogue record for this book is available from the National Library of Australia

ISBN: 9781921833588

Design by Jo Pritchard
Printed by Adams Print, Australia

1 3 5 7 9 10 8 6 4 2

Front cover: The fully mounted skull of *Triceratops horridus* in the *Triceratops: Fate of the Dinosaurs* exhibition space at Melbourne Museum.

Back cover: The mandibular teeth of *Triceratops* were used like scissors to snip vegetation into edible pieces.

JOURNEY OF A TRICERATOPS

EDITED BY CHRIS FLYNN

MUSEUMSVICTORIA
PUBLISHING

We graciously thank our partners and donors for their support.

PRESENTING PARTNER

COLOUR PARTNER

COLOURSMITH®
by TAUBMANS®

MAJOR PARTNER

MAJOR DONORS

SAM AND NINA NARODOWSKI

MEDIA PARTNER

Herald Sun

TOURISM PARTNER

Museums Victoria acknowledges the Wurundjeri and Boon Wurrung peoples of the Kulin Nations where we work, and First Peoples language groups and communities across Victoria and Australia.

Our organisation, in partnership with the First Peoples of Victoria, is working to place First Peoples living cultures and histories at the core of our practice.

Moonrise over Vancouver Harbour and the North Shore Mountains.

CONTENTS

FOREWORD

To stand before something so unlikely as to be almost impossible, is truly awe-inspiring.

Triceratops was an extraordinary beast. One of the largest herbivores ever to walk the planet, it evolved over millions of years to become one of the most successful species of the Cretaceous period. Today, its most distinctive features—the broad frill extending cape-like behind its three horns—appear to us as both familiar and utterly fantastic.

That such a detailed specimen of this remarkable creature has survived tens of millions of years seems almost improbable. Although we don't know how our *Triceratops* died, we do know it remained undisturbed for an inconceivable amount of time—an almost completely intact skeleton, from giant femora to toe bones no larger than a golf ball, encased in what would become the Great Western Plains of Montana. Around it, the world continued to evolve: the Himalayas rose, Australia separated from Antarctica and myriad new species of plants and animals emerged across freshly formed continents.

Today, unearthed and released from the rock, our *Triceratops* is awe-inspiring, something to behold rather than simply look at. And yet as large and intact as it is, this one specimen provides just a small window into the diverse ecosystem of the Cretaceous period. Still ahead is the potential for new discoveries that will reveal in increasing detail the secrets of life on this planet, 67 million years ago.

Perhaps the greatest inspiration the *Triceratops* can offer, however, is to consider our own future. Our remarkable fossil is an unlikely fragment, a chance remnant of a complex and thriving web of life that—save for a handful of descendants—has disappeared. In comparison to the catastrophic event that brought the age of the dinosaurs to an end, our own impact on the environment may feel slight and gradual. But the outcome is the same: what we lose is gone forever, but it's not only chance that will determine our future.

Because it has been acquired for a public collection, visitors to Melbourne Museum and researchers from around the world will continue to learn from Museums Victoria's *Triceratops horridus* for generations to come. I extend my very sincere thanks to the State Government of Victoria and the Premier of Victoria, the Hon Daniel Andrews MP, along with Tim Pallas MP, Treasurer of Victoria, Danny Pearson MP, Minister for Creative Industries and Creative Victoria for this highly significant investment in the state collections.

I gratefully acknowledge donors Vivian Nadir and Susan Narodowski for their support towards the *Triceratops* experience at Melbourne Museum in memory of their parents, Sam and Nina Narodowski. I am also delighted to welcome Coloursmith by Taubmans as Colour Partner for Museums Victoria and *Triceratops: Fate of the Dinosaurs* and thank David Nicholls, PPG Taubmans Commercial Director, Australia.

Major Partner VicHealth (Victorian Health Promotion Foundation) is a much-valued supporter of Museums Victoria and my thanks go to CEO Dr Sandro Demaio for his continued collaboration. Thanks also to our Media Partner Herald Sun and Penny Fowler, Chairman of the Herald & Weekly Times and Editor Sam Weir, and to our Tourism Partner, V/Line and CEO Matt Carrick.

Lastly, my deep gratitude to the Museums Board of Victoria and the huge team of Museums Victoria staff that has delivered the enormous task of bringing our *Triceratops* to Australia with immense skill, professionalism and passion.

Lynley Crosswell
Chief Executive Officer and Director
Museums Victoria

The predentary bone of *Triceratops*, as viewed from below. This forms part of the iconic lower beak.

Some of the fossilised ribs were virtually complete and relatively easy to assemble.

FOREWORD

The Victorian Government is proud to deliver this new world-class attraction and thrilled that Horridus has found a home at Melbourne Museum, where they will be on display for all to enjoy.

The acquisition of Horridus the *Triceratops* into our State Collection marks an epoch in the history of our museum and is a coup for Victoria.

Horridus is the most complete and finely preserved *Triceratops* fossil ever found, and one of the most significant dinosaur discoveries in history.

While this extraordinary fossil dates back 67 million years, its acquisition and display are focused firmly on the future. Horridus is a major new drawcard for Victoria and will introduce a new generation to our museum, attracting visitors from near and far and supporting groundbreaking scientific research in the process.

Having the world's most complete *Triceratops* here means researchers can begin to uncover the secrets of this magnificent species, further cementing Museums Victoria as Australia's leading museum-based palaeontology research program.

The world-first exhibition, *Triceratops: Fate of the Dinosaurs* highlights our unique *Triceratops* fossil in an exciting, immersive exhibit that will inspire awe and curiosity in audiences of all ages.

We will remember a time before and after the arrival of Horridus at Melbourne Museum, whose name and remarkable story will inspire visitors for generations to come.

Danny Pearson MP
Minister for Creative Industries

INTRODUCTION

Dinosaur fossils caused a great deal of confusion when first unearthed in modern times. While researching my book *Mammoth*, I discovered that the Greeks often mistook ancient bones to be those of mythical creatures, such as the gryphon and cyclops. As recently as the 19th century, Europeans struggled to get their heads around the idea of an age of reptiles predating humanity, not to mention the concept of extinction. As palaeontologists strove to piece together fragments of fossilised bone, their work was ridiculed as fantastical speculation and, in some cases, heresy. If only those pioneers of science could see *Triceratops horridus*, the most complete specimen ever recovered, on display in Melbourne Museum.

Transporting a dinosaur from the United States of America to Australia during a global pandemic proved a tad more difficult than buying a pair of shoes on eBay, but thanks to a disparate and determined range of people, Horridus embarked on an incredible modern journey that began 67 million years ago.

In these pages we hear from Craig Pfister, who discovered the specimen in the Hell Creek Formation, Montana, as well as Terry Ciotka and his crew at Dino Lab in Canada, who performed magic by cleaning away the rock surrounding the fossils, casting a steel armature and packing the bones into crates for their long journey to Australia.

On the Australian end, First Peoples collections manager Nancy Ladas was there to welcome Horridus to their new home, while palaeontologists Erich Fitzgerald, Tim Ziegler, Hazel Richards and a host of dedicated Museums Victoria staff painstakingly unpacked, catalogued, scanned and made preparations to exhibit our new friend to the public. Their excitement, wonder and curiosity illuminate these pages as brightly as the stunning photographs that chronicle every step of the journey.

Nothing like this has occurred in Australia before. We have a real dinosaur on display in Melbourne. One that lay down, fell asleep for a few eons, and woke up in our museum. It's so achingly beautiful and whole—it feels alive. Experiencing this fossil inspires a sense of awe, while also urging reflection on loss and extinction. But perhaps the most powerful and unexpected feeling is one of connection.

This is apparent in the reactions of many other contributors. Shaun Tan tells us about how his own artistic journey has been inspired by *Triceratops*. Shannon Martinez creates a prehistoric menu. Benjamin Law wants to lay on its belly. Legendary scientists like John Long, Philip J. Currie, Susie Maidment, Tim Flannery, Tom Rich, Linden Ashcroft and Dermot Henry explore the aspects of *Triceratops* that enthral and fascinate in their own fields. Artists, curators and intellectuals such as Rebecca Dart, Tony Birch, Maree Clarke, Laura Jean McKay, Mark Radburn, Mitch Mahoney and Shannon Faulkhead talk of memory, nature, beauty and love. This one creature means so much to so many.

Journeys have beginnings and endings, notionally, but nature and time keep going. They don't stop. Life finds a way. This *Triceratops* is here, it's real, it's still with us, it's unique. It has endured, just as we will endure. A *Triceratops* in Melbourne. It was always meant to be. Horridus has been slumbering all this time, waiting for us. We were always connected. This city, our people, First Peoples, Melbournians, Victorians, Australians, Canadians, Americans, Earthlings. We all come from the same place, and we all wind up in the same place. Our *Triceratops* is a reminder that we share a common bond. That we are here only briefly, and yet are never truly gone.

Chris Flynn
Editor-in-residence, Museums Victoria

Exhibition Collection Manager Anthony Abell checks the position of the skull, which has been bandaged for protection during the installation.

Horridus's skull begins to take form.

Shannon Faulkhead

Head, First Peoples Department
Museums Victoria

The first time I went to the museum and actually saw mega fauna was incredible. As kids we heard stories about giant wombats and kangaroos. One of the language animations I worked on at Monash University was about a giant lizard, so these big animals are very much part of our history and the stories that we are told.

Measuring importance based on age is an odd concept. This *Triceratops* exists. They may not be living but they still exist. Their stories are alive. When I say 'stories' I don't just mean the stories of them roaming across the plains, but also the stories that were shaped during the excavation where people mistook bones for something else, almost creating whole new animals within the dinosaur. Just because they're no longer alive doesn't mean they don't exist anymore, so how do you think of them as being something historic?

Within the First Peoples team, when we talk about ancestors we don't just mean bones or human material. We talk about how a person's essence is still part of an item that they made to be used every day of their lives. Everything in our collection has a connection to somebody or somewhere. It's not unusual for us to talk to them. Admittedly, if we're in the hallway we tend to do it silently. We don't think of a lot of the collection as inanimate. It's still living. It has life.

Working at the Koori Heritage Trust, an Elder would pull an item out of the collection and start telling you about it as if the person who made it was there. It's nice to think of the museum as a place where everything is with us still and can speak to us. It's that concept of continuum. Life is one big circle, with everything coexisting and living in different forms.

Horridus will ignite the imagination of visitors, understanding that those little toys you got as a kid were in reality bigger than a house. It's not just the size of the specimen—it's also the sociological imagination, the ability to view the personal in a broader historical context, and the resulting emotions that will be stirred.

Will Horridus feel homesick? Will they want to be back there? Not that 'there' exists in the same way anymore. The past is never gone. It's still here. Maybe you just can't see it, or don't look closely enough.

The predentary bone, located at the tip of the lower jaw, forms part of *Triceratops*'s distinctive beak.

The skull of Horridus is 99% complete. Pictured is the right side of the jaw and mandibular teeth.

The discovery site lay deep
within the Montana wilderness.

> The past is never gone.
> It's still here.
> Maybe you just can't see it,
> or don't look closely enough.
>
> Shannon Faulkhead

Craig Pfister

Palaeontologist

Finding this fossil was a wonderfully unexpected discovery, and I feel most fortunate and privileged to be part of its journey. When I first discovered the *Triceratops*, I thought it had the potential to be a relatively good specimen. After excavating for three to four weeks I could see that it was fairly intact, and by the end of the first field season I knew it was exceptional. It is my hope that this *Triceratops* will always be available for public viewing and scientific study.

If *Triceratops* were alive today, I suspect they would find our current era pleasant enough, although a bit chilly and terribly overrun with mammals.

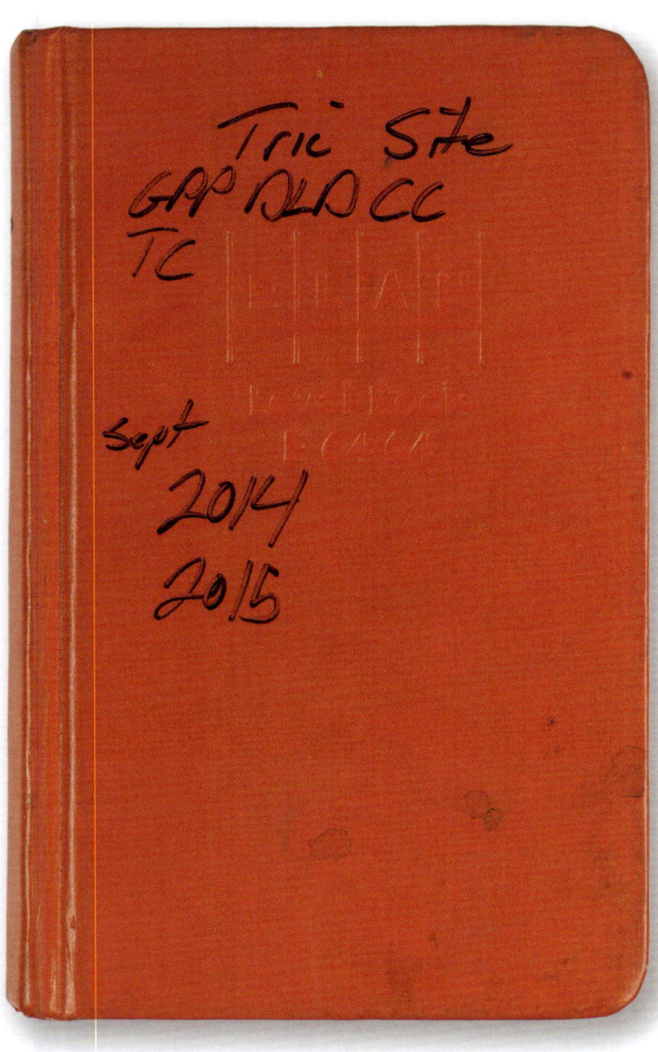

Craig Pfister worked tirelessly in hot conditions to excavate Horridus from the sandstone of Montana, recording his progress in a field journal (above).

As the stone was carefully removed, the import of the discovery in the Hell Creek Formation became increasingly apparent.

The Hell Creek Formation, Montana.

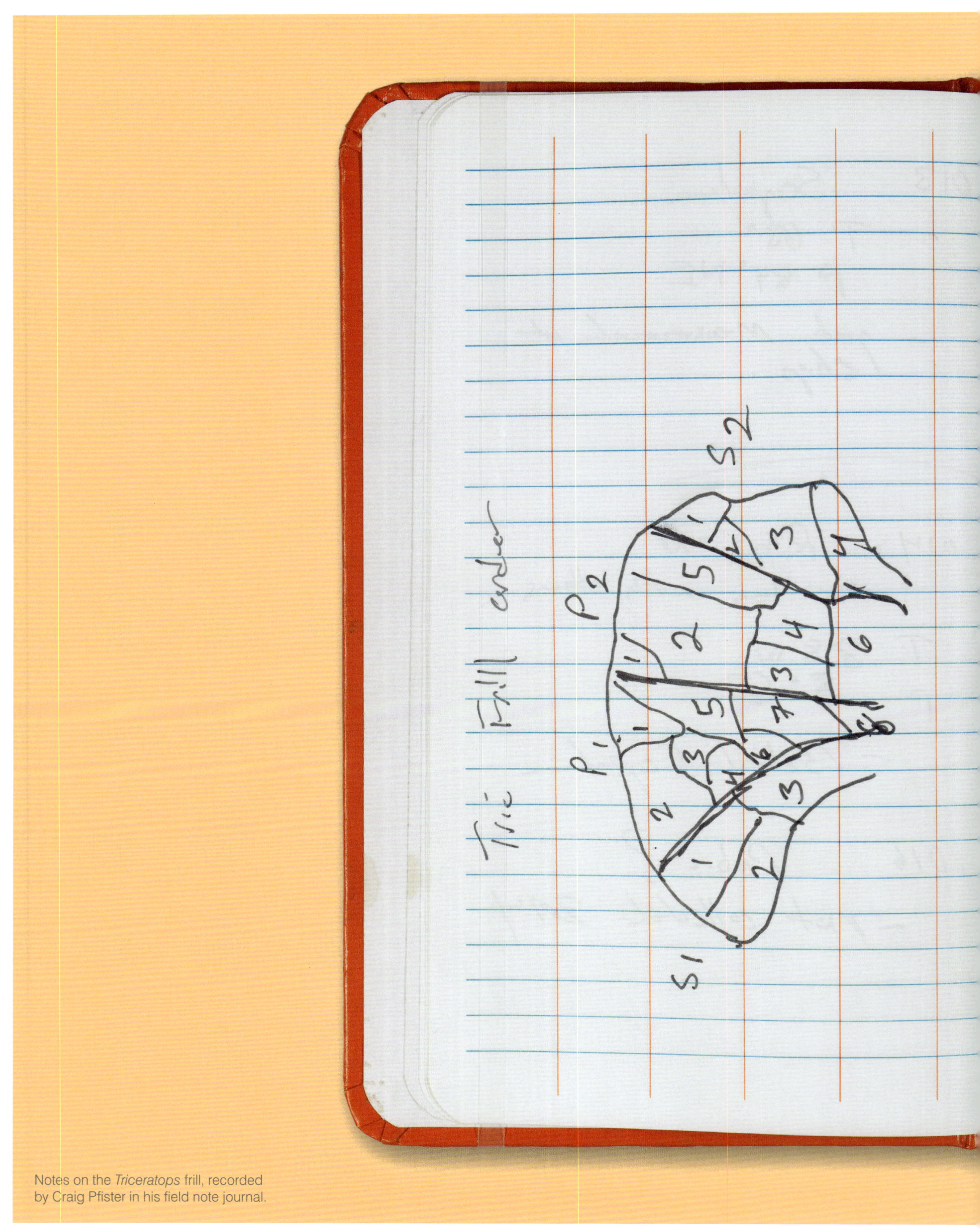

Notes on the *Triceratops* frill, recorded by Craig Pfister in his field note journal.

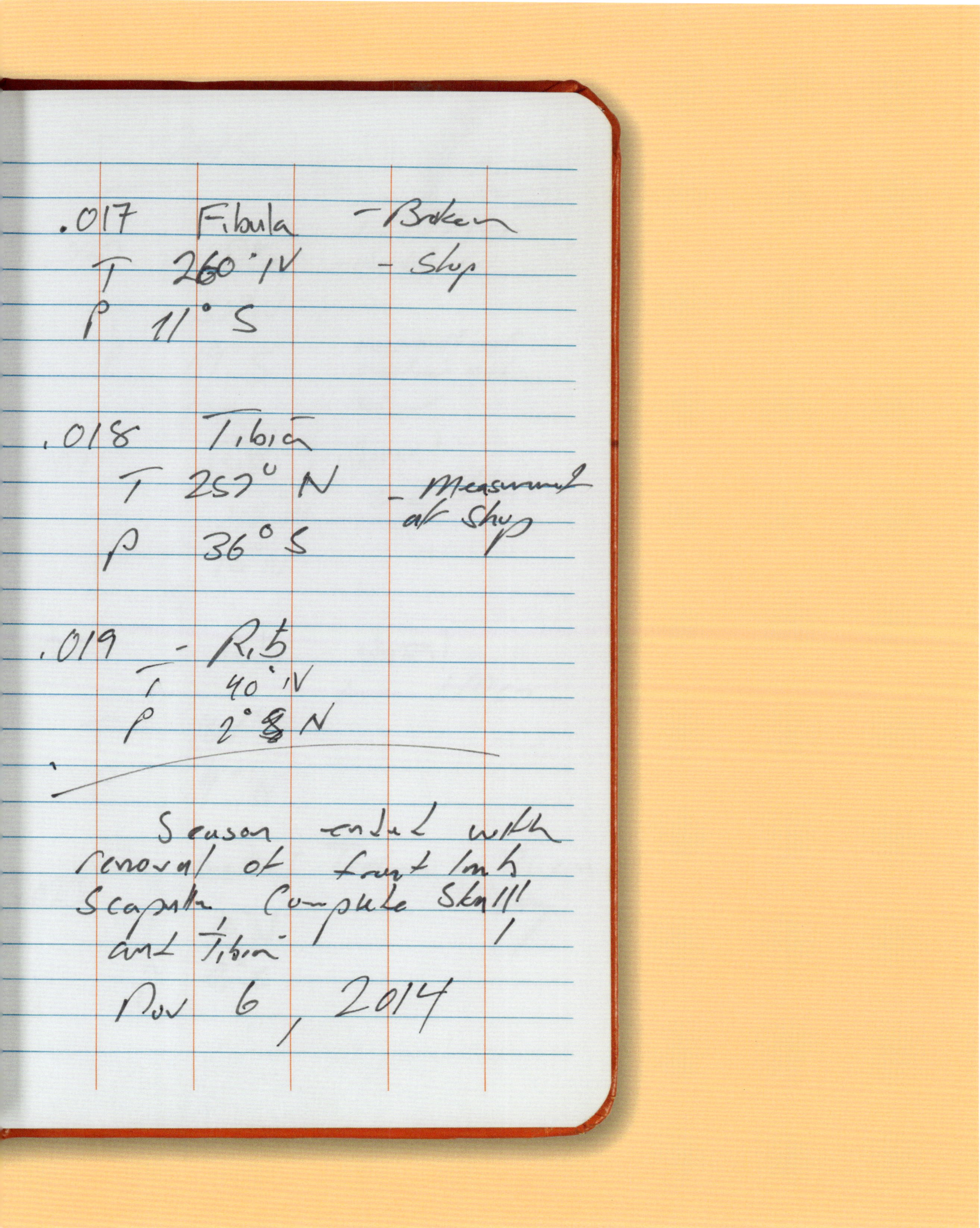

.017 Fibula – Broken
T 260° N – Slip
P 11° S

.018 Tibia
T 257° N – Measurement at Slip
P 36° S

.019 – Rib
T 40° N
P 2° 8 N

Season ended with removal of front limb Scapula, Complete Skull and Tibia

Nov 6, 2014

The fossilised head of Horridus is gradually uncovered by Craig Pfister in the Hell Creek Formation, Montana.

These tendons were already ossified when the animal was alive—this is common in this group of dinosaurs. Because they are mineralised in life, they are preserved like other bones when the soft tissues are lost prior to fossilisation.

Erich M. G. Fitzgerald

Senior Curator of Vertebrate Palaeontology
Museums Victoria

Triceratops is a well-known dinosaur and perhaps because of that people haven't been given the opportunity to gain insight into what is strange and weird about it. What's behind the frill and horns of *T. rex*'s archnemesis? That's the real story.

We can glean information from this unique specimen that will enable us to tell new stories about the history of our planet, life on Earth, dinosaurs and of course *Triceratops* itself.

When I was a kid I went to the National Museum of Victoria, in the building which is now wholly occupied by the State Library of Victoria. I remember walking into the old McCoy Hall and seeing replica skeletons of dinosaurs. They had only a few. As a child, I pored over books featuring lavish photographs of the dinosaur displays in the great museums of the North Atlantic axis. One day I hoped to see those real bones. I wondered why there weren't real ones here. I knew there were dinosaurs from Australia, but those specimens were far more fragmentary. The geology of Australia wasn't conducive to producing fossils like the ones in those books I read. Imagine the impact of seeing the real thing—learning Horridus's story and being able to come back and see the exhibit repeatedly, knowing it's not temporary. That has power, and provides value to society and education that is hard to quantify. That's where the impact will really lie.

For me, potentially the most rewarding aspect of this acquisition is knowing that this opportunity to experience the real thing is here.

I think people are going to be awestruck by Horridus. It's difficult to grasp how incredible this skeleton will be as a spectacle. I had quite an emotional response when I saw it all assembled for the first time.

Some will look at this *Triceratops* and ask, 'How did that happen? Why did these creatures become extinct? How did the universe or nature allow this to occur?' It's unsettling. It raises uncomfortable existential questions around how and why extinction events are happening now.

One thing that struck me about *Triceratops* was how little we know about the skeleton behind the skull. When we first announced the acquisition of Horridus to the media, we made a big deal about the fact we have its tail. The response was, 'The tail? What about the skull? You've got the wrong end, mate.' But think about how little we know. *Triceratops* was first described only 130 years ago. We never knew, until now, how long the tail was. It's quite a bit longer than any piece of skeletal reconstruction or art depicts.

In Horridus, we have almost all the bones from one individual, unambiguously found together in articulation. It's a real, formerly living biological unit, as opposed to a chimera of many different individuals. This is a canonical example that can be used for comparison with other horned dinosaur fossils and indeed other species beyond *Ceratopsidae*. That has huge scientific importance.

Given the abundance of *Triceratops* in the ecosystems at the end of the Cretaceous period, if you can understand the paleobiology of *Triceratops* at a finer grain then you can gain insights into how its ecosystem was working just before the end of that era. That's the most profound scientific potential of a fossil like this. It's beyond *Triceratops* itself. Ironically, the associated circumstantial evidence gleaned from the bits and pieces around the *Triceratops* may have a longer lasting significance than the bones themselves, by revealing what role *Triceratops* played in its environment. *Triceratops* was the most abundant herbivore in its ecosystem, and understanding what it might have been eating, its locomotion and how fast it grew leads into population dynamics and demographics of this species. We may even be able to get a glimpse into its health, prior to the cataclysm 67 million years ago.

The skull gives us an indication of the size and weight of *Triceratops*.

An armature was created by the team at Dino Lab in Canada, specifically to house the fully articulated tail.

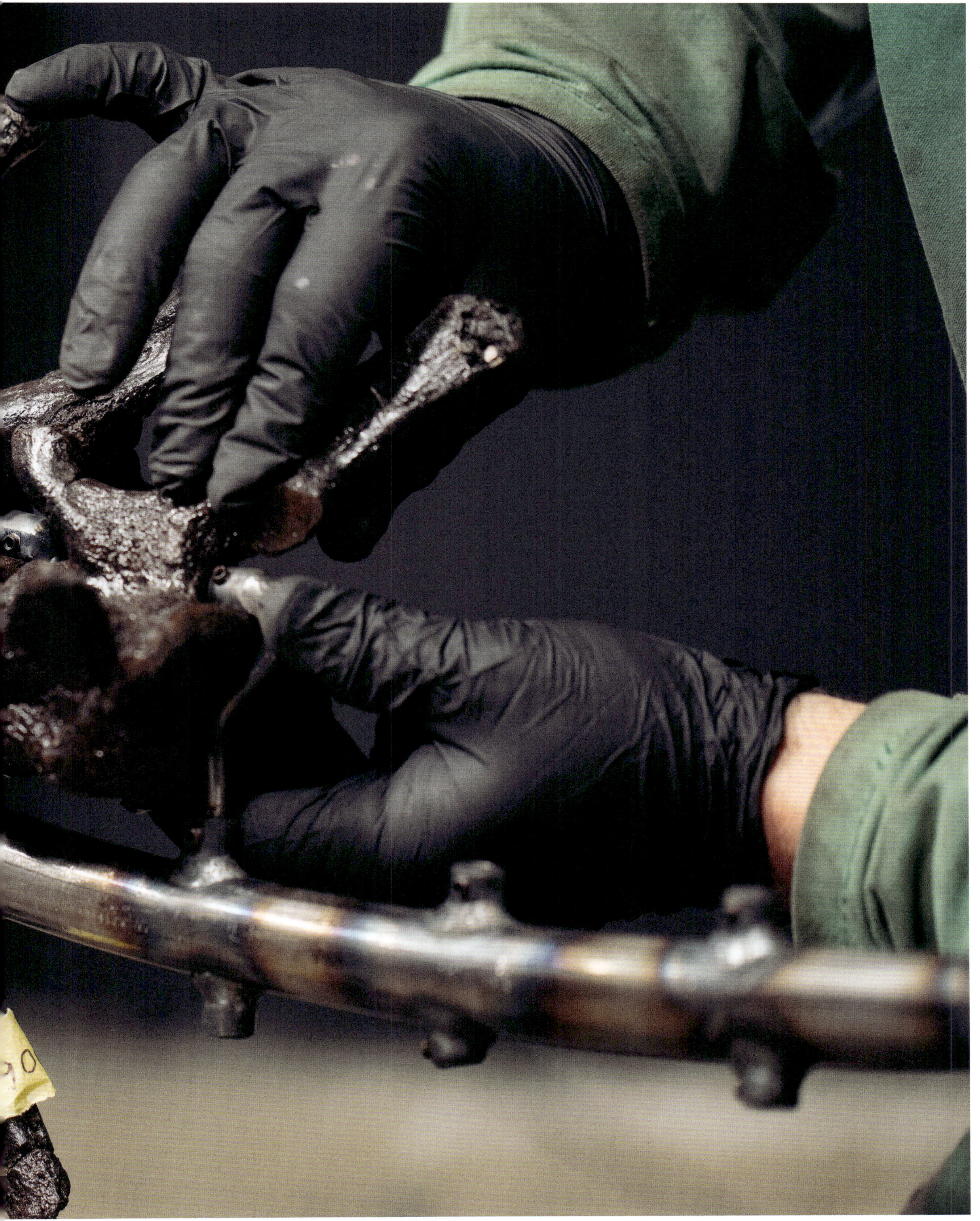

Thinking about if *Triceratops* were alive today, there's no analogue. Describing it as an elephant-sized thing with a long tail, a parrot-like beak and really big horns doesn't even do it justice. If it were alive today, you might have to keep it indoors, in a hermetic greenhouse biosphere environment. The lost-world fantasy-island theme park scenario wouldn't work. *Triceratops* doesn't belong anymore. Its environment was so different to anywhere that exists today.

If adults can for just a moment be taken back in time to their childhood then all the hard work to bring Horridus to life will be worth it. I'm a firm advocate for the adventure of discovery. The pay-off, in addition to the scientific discoveries, will be getting people to think about horizons again. There's a lot to deal with on our planet on an environmental, technical, political, social and emotional level, but there is also so much undiscovered. Giving people the opportunity to experience the thrill of seeing Horridus in Melbourne, of all places, is so exciting.

Museums Victoria's Erich Fitzgerald visited Vancouver Island in order to document the specimen.

Miller
Museums Victoria

It's a real, formerly living biological unit, as opposed to a chimera of many different individuals.

Erich M. G. Fitzgerald

The sacrum of this *Triceratops* specimen was remarkably intact, complete with ossified tendons.

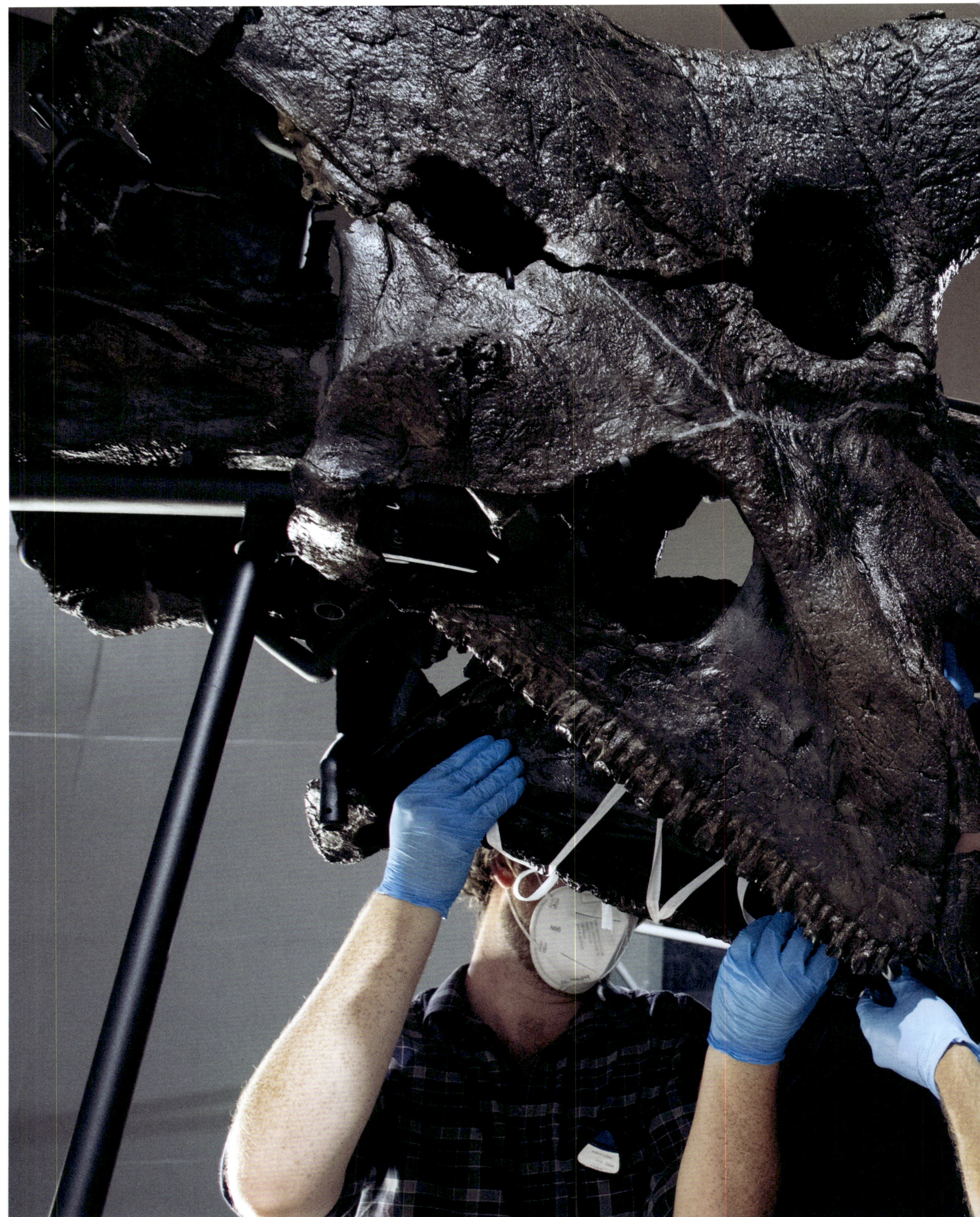

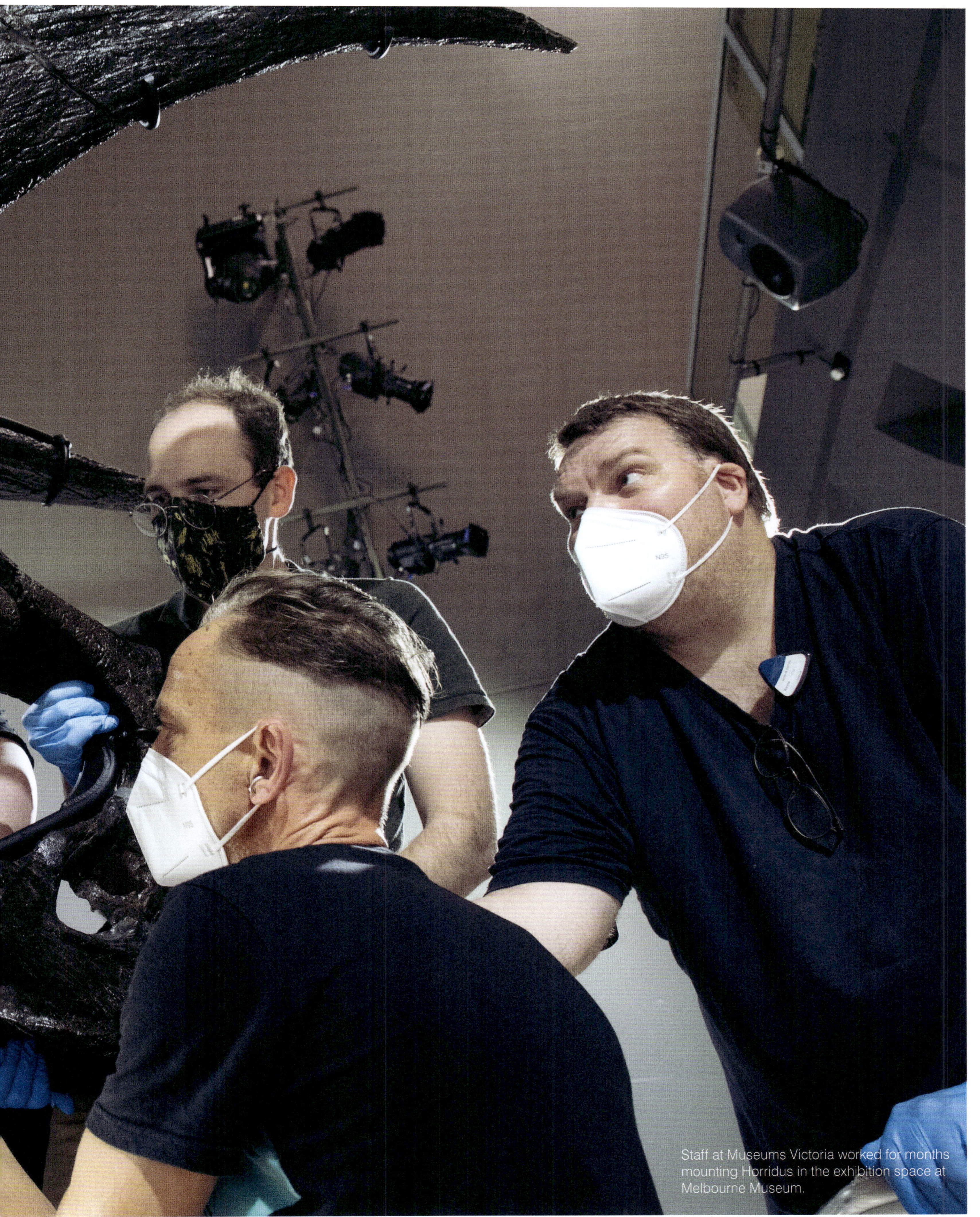

Staff at Museums Victoria worked for months mounting Horridus in the exhibition space at Melbourne Museum.

Melissa Kay

Fossil Restoration Technician
Dino Lab

I did a lot of finishing on the bones of Horridus. After you take the matrix off the bone you still have to get in there and ensure it's all clean. In the case of Horridus, that was easy. The bone was a completely different colour from the matrix, which mostly fell away easily. It was really beautiful and easy to work with. I get to see parts of the bone that other people wouldn't, because once the dinosaur is mounted, you're not going to look too closely at certain areas.

My formal background is Fine Arts. I find that clearing away the bone underneath the matrix is exactly the same as drawing. It's the same process of sitting down, paying attention to detail, getting completely absorbed in the act of doing it. And it requires patience. If it doesn't look right, you might have to go over the same spot 50 times, and that's okay. You let the bone tell you what it wants.

Melissa Kay working on the fossil.

This sequence of dorsal vertebrae sit immediately in front of the sacrum, something like the 'lower back' region in humans. These vertebrae had so many ossified tendons preserved intact on the top side (underside here, not yet discovered as they are still embedded in the rock in this picture) it was decided to leave them articulated in a single block of seven vertebrae.

Surgical tools are used when cleaning away rock from fossilised bones.

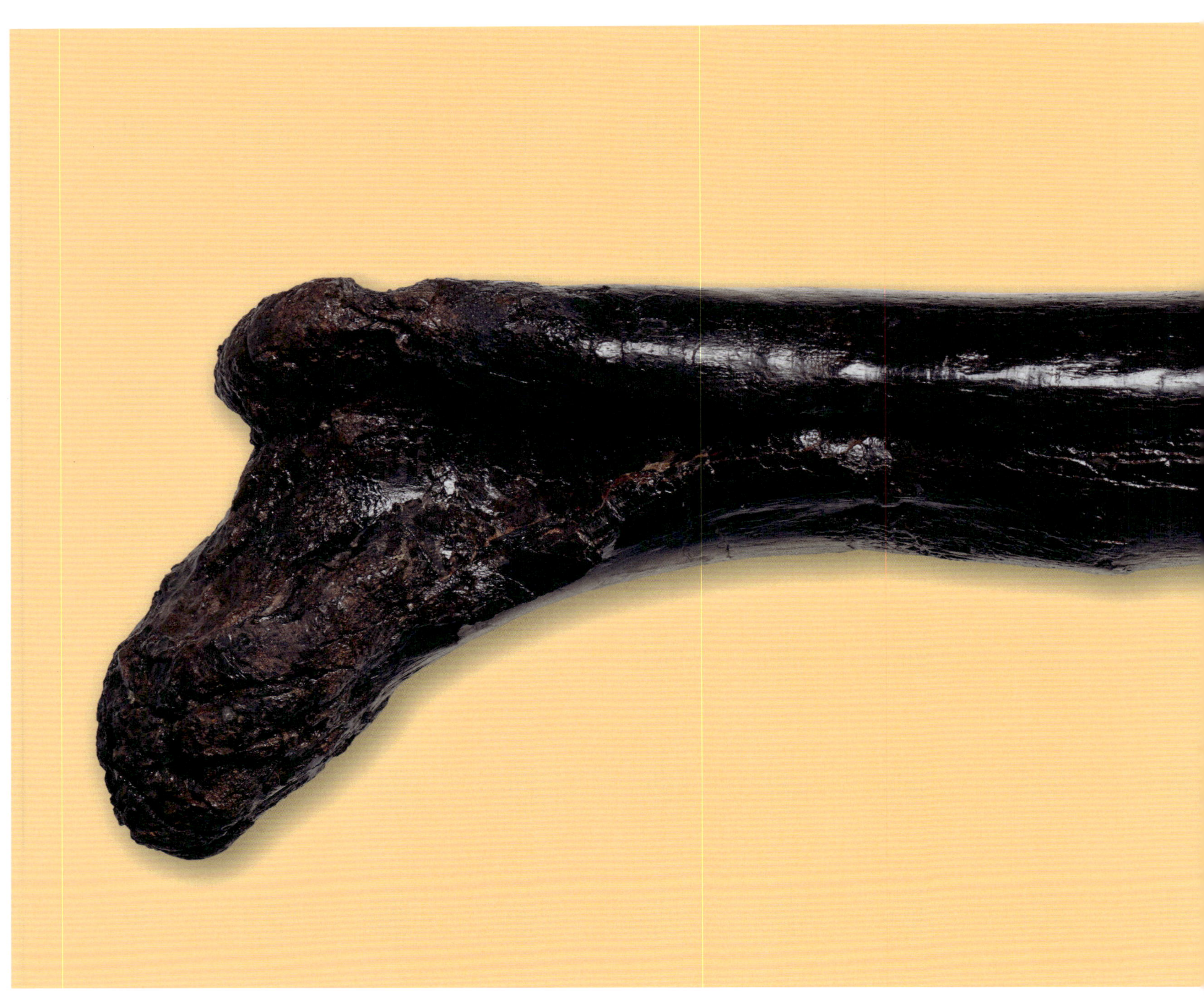

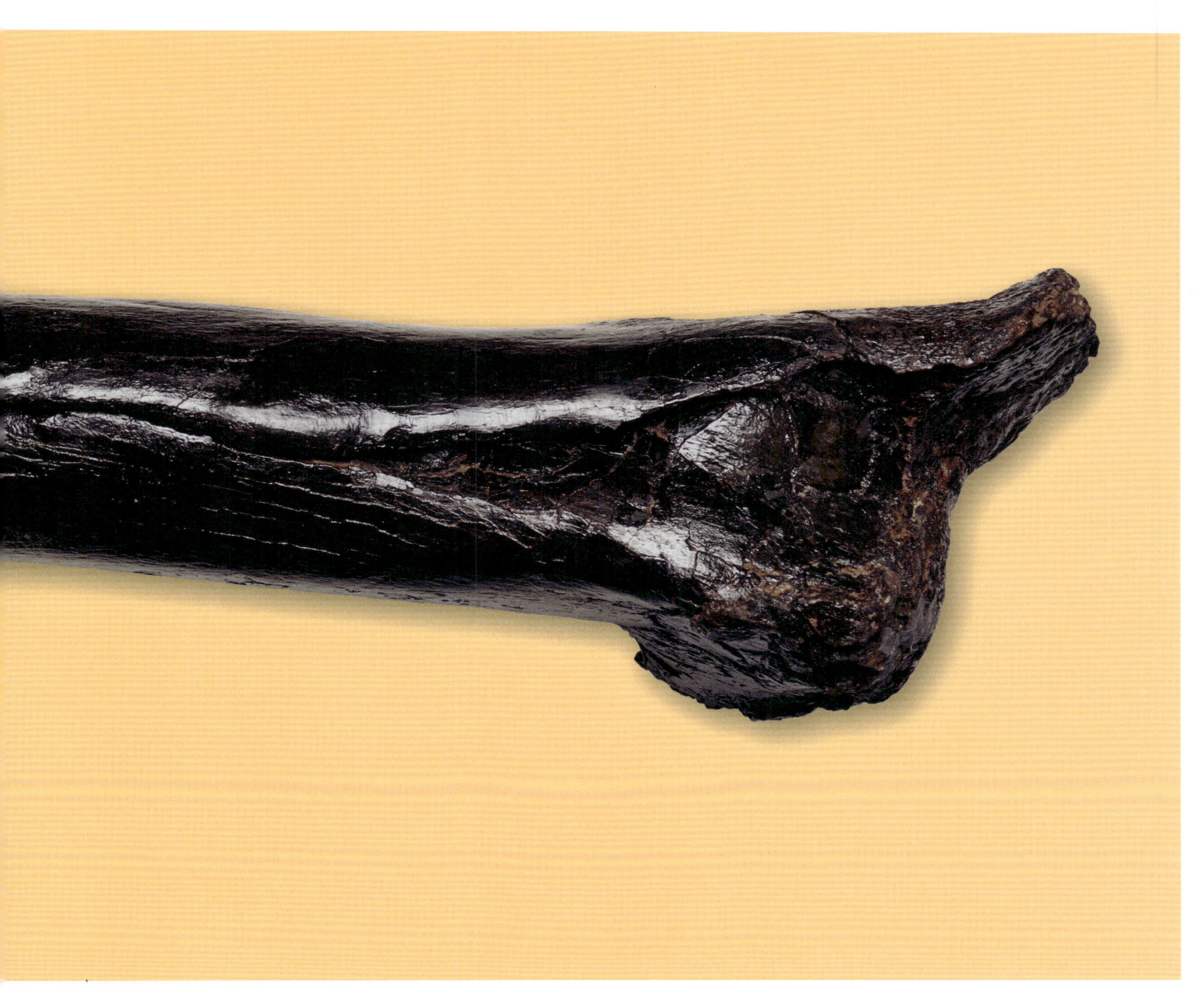

The left femur is one of the longest individual bones in *Triceratops*.

Removing the fossil from rock involved months of intricate work using surgical tools.

> If it doesn't look right, you might have to go over the same spot 50 times, and that's okay. You let the bone tell you what it wants.
>
> Melissa Kay

Tim Flannery

Palaeontologist + Environmentalist

I pestered my mum into buying far too many packets of Weet-Bix purely for the plastic dinosaurs that came in each box. My youngest son loves the ceratopsian family. All those crazy horns and names. *Kosmoceratops, Xenoceratops, Triceratops*—how rich and wonderful. There's a whole dinosaurian biography in this skeleton.

The joints between the vertebrae were preserved in complete articulation, and the heads of the ribs are almost exactly where they were when the animal was alive. This picture represents the view from inside the ribcage, looking up.

A portion of the jaw highlights the mandibular teeth *Triceratops* used to slice through flora.

3A - String
of Dorsal

As bones were removed from surrounding rock, they were identified and individually labelled.

Benjamin Law

Author + Journalist

I was a massive dinosaur nut as a kid. My two superpowers growing up were the ability to spit out and understand every Simpsons reference and knowing everything about dinosaurs. I knew their names and family groups. I was a huge fan of sauropods. I can't remember a time when I wasn't into dinosaurs. I was eleven years old when *Jurassic Park* came out. The ground-breaking technology that made the dinosaurs in the film look real tipped me over the edge into complete obsession.

At that age, when we were absorbing language, it was wild that as kids we could reel off words like *Brachiosaurus*, *Archaeopteryx* and *Tyrannosaurus rex*—they almost sounded like incantations. The fact that these monsters existed is magical.

Triceratops was one of my favourite dinosaurs because I loved herbivores. Probably because I watched *The Land Before Time* and herbivores were the cute ones. *Triceratops* is such a noble creature. I love rhinos, hippos and elephants—creatures with a sense of nobility, not designed to kill because they're herbivores, but obviously able to defend themselves against an enemy. I love the pageantry of the *Triceratops*, its theatricality. The crest is pure drag queen.

This seemingly unremarkable lump of bone is symbolic of the amazing preservation of the specimen. Horridus was so intact that even tiny bones, like this one from the wrist (carpal), were preserved in situ.

Triceratops is one of the most instantly recognisable creatures that ever lived.

The caudal vertebrae form part of *Triceratops*'s tail, which has never been found in such complete articulation before.

A conspiracy of geology has left this evidence behind for us. Not every skeleton becomes a fossil, so there's a sense of gratitude and wonder that this fossil even exists, that we have this ancient record of a creature and how they lived.

I've always been captivated by that scene in *Jurassic Park* where you see a living *Triceratops* that has accidentally eaten something it shouldn't have. I could happily lay down on the belly of a *Triceratops*. That's a lovely daydream.

The frill is held up by an intricate network of steel supports.

> ‘The crest is pure drag queen.
>
> Benjamin Law

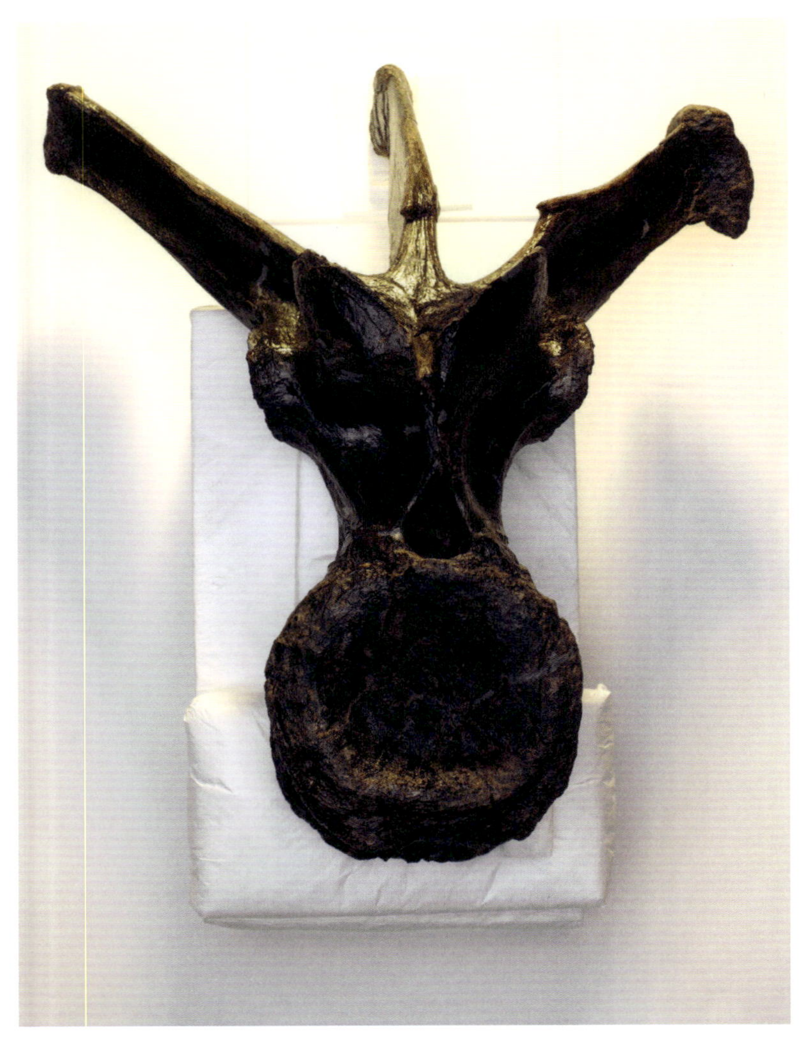

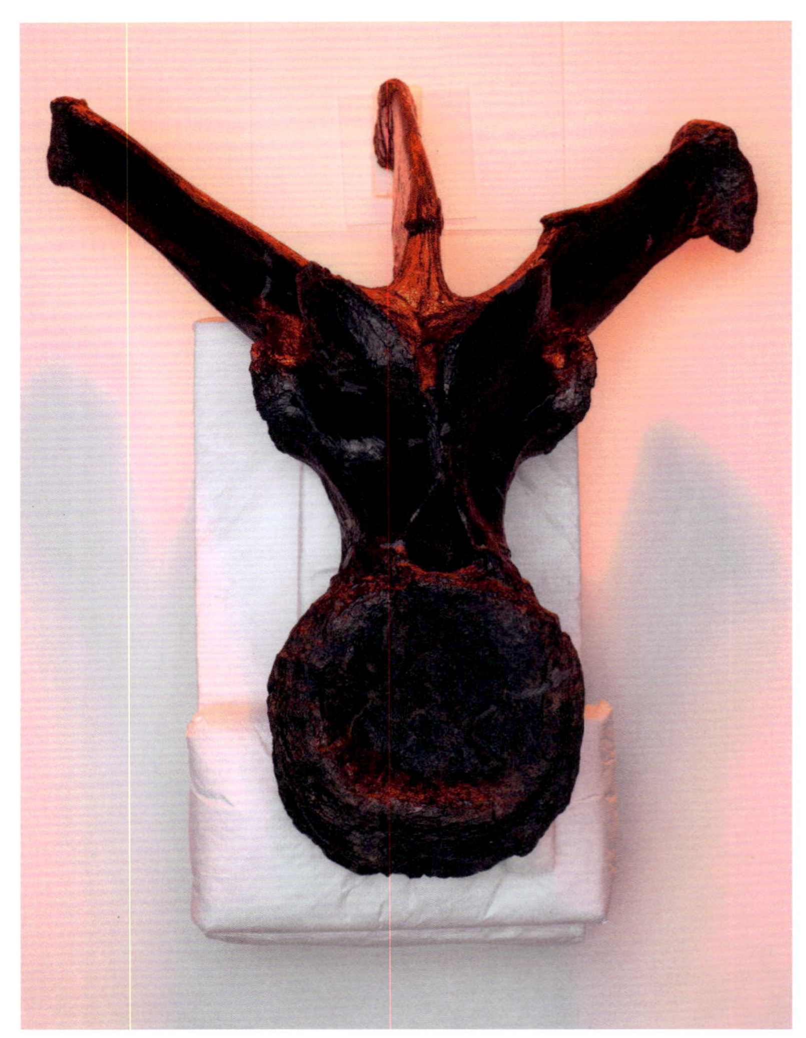

Maree Clarke

Curator + Artist

I see beauty in everything. I look at jawbones and I see jewellery. I think about what else we can do with those bones and teeth. How they can be repurposed as beautiful body adornments that people can wear. This is my traditional cultural practice made contemporary. I'm giving animals another life. They haven't died in vain.

My favourite part of the body is the teeth and jaw, but I can see how the *Triceratops* tail could be turned into a beautiful 3D printed necklace. Usually when making my jewellery I have to supersize, but in this case I'd have to scale it down!

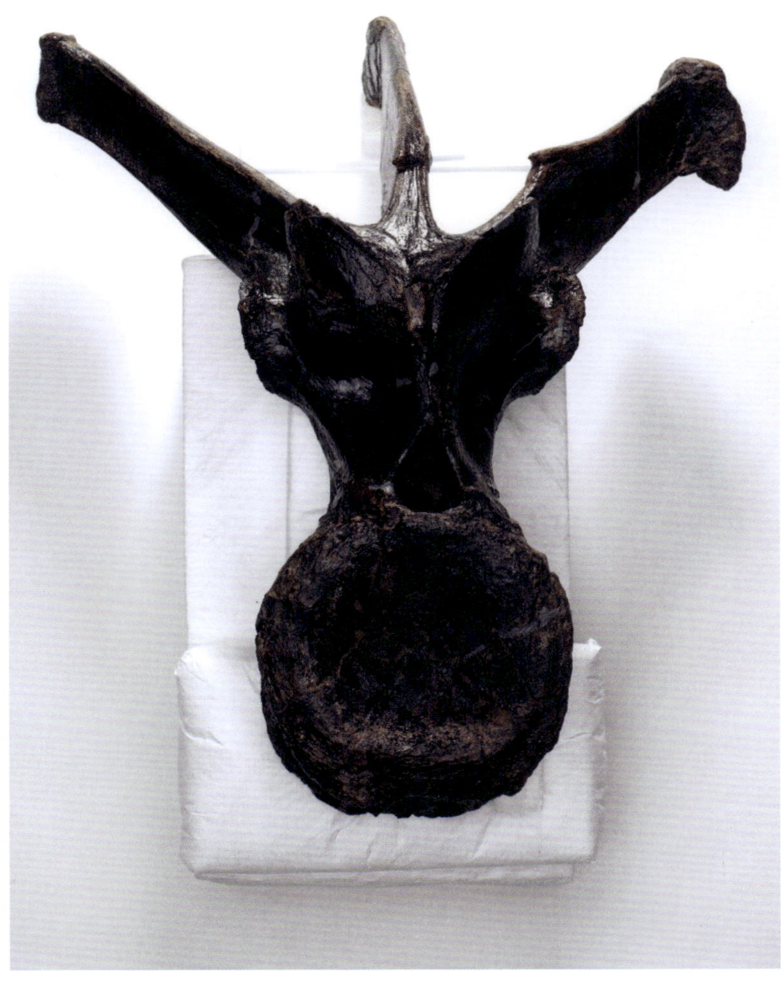

The dorsal vertebrae begin behind *Triceratops*'s neck and run all the way to the sacrum at the top of the tail, forming the incredibly complete spine of this specimen.

The tip of the tail is a point where function and design meet.

> My favourite part of the body is the teeth and jaw, but I can see how the *Triceratops* tail could be turned into a beautiful 3D printed necklace.
>
> Maree Clarke

> I see beauty in everything.
> I look at jawbones and I see jewellery.
>
> Maree Clarke

The mandible combines with the predentary bone to evoke the memory of a very real, living creature.

Professor John A. Long

Flinders University

I grew up collecting fossils around Victoria, and finding a dinosaur was always my dream. As a kid, fossils were my entire world. I loved reading about dinosaurs, drawing dinosaurs and even writing school essays about dinosaurs. Dinosaurs are the first introduction to science for many kids, and I knew when I was seven that I would become a palaeontologist when I grew up. Now I am, and I have found dinosaurs and even named a new species (including Australia's oldest predatory dinosaur, *Ozraptor*).

Triceratops was my favourite amongst the plant-eaters. I still recall the breathtaking 1927 Charles Knight mural of *Triceratops* standing in the mists of time, boldly challenging the fierce *Tyrannosaurus*, head raised, horns poised. *Triceratops* was loaded with dignity and had a charisma about it—a grazing beast that just wanted to get on with life yet was always threatened by gargantuan predators. Somehow it survived. Thrived, in fact, if the numbers of bones found in American late Cretaceous deposits are anything to go by.

To see real bones and imagine them covered in flesh is powerful. The skeleton is a time capsule of knowledge holding many layers of scientific information. You just need the right tools to retrieve this data. It's highly likely that one day Horridus will reveal something about the lost world of dinosaurs that no-one else has ever known.

We can make 3D scans of the skeleton to test how it walked or ran. We can study its teeth to determine what kinds of plants it ate and how it chewed. We can study its skeleton for pathologies or wounds to see if it lived a healthy life and discover what kinds of encounters it had with other dinosaurs. The bones can tell us the body temperature of the dinosaur when it was alive. The chemistry of the rocks the fossils are found in can be used to calculate ancient temperatures and environmental conditions, including the nature of the terrain the dinosaur lived in. The fact that this dinosaur lived more than 67 million years ago tell us that the diversity of life on Earth has continually changed over time, as have the movements of the continents and the climate of the world.

Dino Lab staff spent hundreds of hours perfecting the armature so it would support the weight of the bones.

"*Triceratops* is aesthetically more beautiful than any other ceratopsian because of its symmetry.

Professor John A. Long

Triceratops ('three-horned face') is renowned amongst dinosaurs for its iconic horns. Even after 67 million years, they still inspire a sense of awe and wonder.

Dinosaurs ruled the Earth for more than 160 million years, but all of them (except the bird lineage) became extinct about 65 million years ago, dramatically changing the nature of animal life on Earth. All life on Earth is in flux. Small, opportunistic creatures often survive mass extinctions. Birds, small reptiles and mammals survived the KT asteroid strike, while the largest dinosaurs and marine reptiles became extinct. Are we now amid the sixth mass extinction event? I think we are, due to human-induced environmental changes wiping out many species each year.

If Horridus were alive today, they would love the Otways and the Strzelecki rainforests of Victoria. Healesville, for sure. Grasses had not evolved at the time *Triceratops* lived— it relied on ferns, mosses and conifers, which grow in abundance in the rainy mountainous regions just outside of Melbourne.

Triceratops is aesthetically more beautiful than any other ceratopsian because of its symmetry.

The tail of Horridus was found in full articulation at the dig site in Montana.

It's highly likely that one day the Museums Victoria *Triceratops* will reveal something about the lost world of dinosaurs that no-one else has ever known.
Professor John A. Long

NMV

The vomer is a narrow bone of the front part of the palate which separates the nasal cavities. Its name derives from the Latin word for 'ploughshare'.

Terry Ciotka

Pangea Fossils

When I'm hiring staff at Dino Lab, I try to find people who have attention to detail, who are good at 3D puzzles. They also need to have artistic ability. I never find people with experience in this field, because they don't exist. Cleaning a single bone can take a month, sometimes even longer, especially with the Melbourne *Triceratops*, because we did almost everything with a knife, gently scraping and watching for cracks in the surrounding matrix. It's a very lengthy and hard process, but we love it. Dino Lab is a great place to work.

We create an armature that goes around the bone, but everything can be taken off and it fits back on perfectly. With a lot of dinosaur bones I use magnets. We attach a magnet to the bone using a special glue which can be dissolved without damaging the bone. With a few of the smaller dinosaurs that we've done there's virtually no metal visible—the dinosaur looks like it's just standing there. It's going to potentially change the face of palaeontology, because they look so sleek and cool.

The armature is an incredible piece of functional art. A lot of times it looks just like the dinosaur. When we were putting the skull together, myself and one of my preparers, Melissa, had to hold the frill in place while the armature was being welded. We were holding it for thirty minutes and our arms were shaking. People don't see that part of the process, but if you ask any one of us here, we remember that day vividly. When you're holding a *Triceratops* frill for half an hour until it's in place and then you stand back, it blows your mind. There are really no words to express how it feels, because you're one of a handful of people that have ever been able to do this. When we're cleaning a bone and uncovering it for the first time, it hits you that you're the first human on the planet to ever see this 67-million-year-old creature.

I've always liked dinosaurs, but I never thought I could make this a career. I just got lucky.

When we're preparing bones, the whole time we're imagining what the animal's life was like when they were alive. Sometimes you find something unique within a bone. Maybe there's a weird hole or a bulge in the bone, which means the animal hurt itself in life and was healing. The Melbourne *Triceratops* has a couple of injuries. We can theorise as to what happened. Did it fight with a *T. rex*? Did it happen while mating with another *Triceratops*? When you're working on the bones, your mind wanders. You take yourself back to the Cretaceous and it's almost like sitting with the dinosaur, watching it live.

We have bones that I have never seen before. Everything we had was articulated. *Triceratops* often died together, so it's common for their bones to be jumbled up at dig sites. There are only seven *Triceratops* in the world that are not composites. Having something this complete and isolated gives us a good indication of its environment. There was a weird section of organic matter in the stomach area. We were convinced it was stomach contents, but it could've been residue from the river that slowly washed up and settled in the stomach. One of the leading researchers in the world was able to identify that there were pollens and seeds in there, so we know it's organic matter. Whether it was their last meal or not, we may never know. One palaeontologist believes it died on the edge of a river and was slowly covered, which is why none of the bones really moved—it's perfectly articulated in the position in which it died. I've never seen anything like it before.

When we finished, there were a few tears. To see this magnificent dinosaur standing in front of us after a year and a half's work, we got a sense of appreciation for how beautiful it was. Philip Currie, one of the most recognised palaeontologists in the world, said this was not only the best dinosaur he'd ever seen but the best documented. Thousands of photographs, video, scans of every bone. We did so much preservation. We even kept the soil. Scientists will have so much data to study. It will be wonderful seeing what they discover.

The steel armature was a bespoke design, sculpted specifically for Melbourne's *Triceratops*.

Extreme care was taken while sculpting the armature to fit the shape of the fossil.

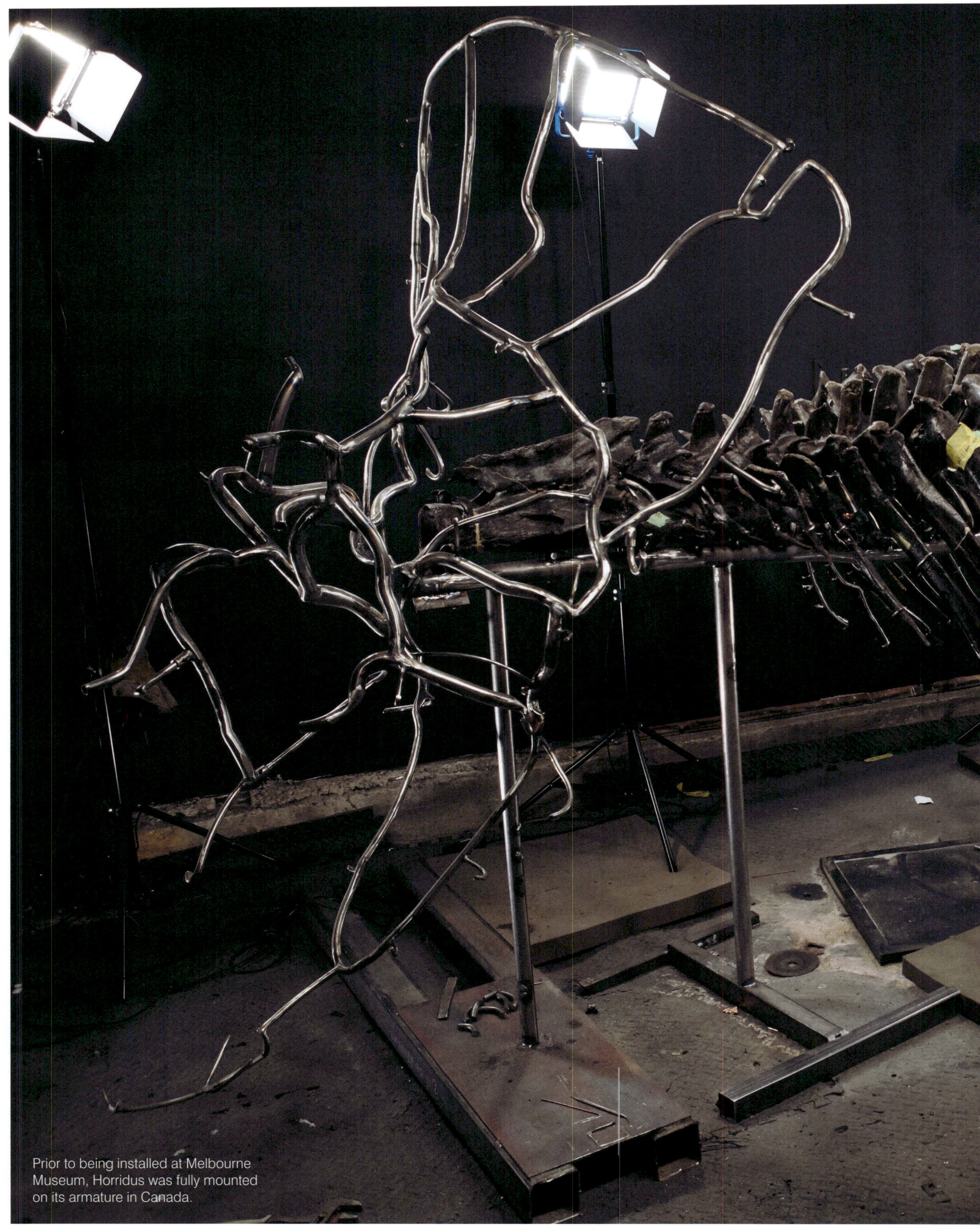

Prior to being installed at Melbourne Museum, Horridus was fully mounted on its armature in Canada.

> The armature is an incredible piece of functional art.
>
> Terry Ciotka

Unusual shapes proved a challenge for the experienced welders.

> When we're cleaning a bone and uncovering it for the first time, it hits you that you're the first human on the planet to ever see this 67-million-year-old creature.
>
> Terry Ciotka

The fossil was carefully cleaned by preparators.

Rebecca Dart

Art Director, Designer, Animator

I inherited a bunch of old and worn children's dinosaur books from the early 1960s from my older brothers. They depicted dinosaurs as slow, lumbering tail-draggers, but as a child of the 80s I started seeing paleoart of lively, fast-moving animals with interactive behaviours. It was exciting to see such a fine interlocking of science and art.

The *Triceratops* exhibit at Melbourne Museum showcases a complete ecosystem, from giant dinosaurs to tiny mayflies. Some animals are unique to that time and others are familiar to us today. It informs us not only of what the animals could have looked like, but how they could have moved and acted. It envelops us with the plants, animals and environments of the late Cretaceous. Who wouldn't want to go there for a while?

When you see Horridus you're struck by the scale—by the sheer size of the animal—and by its uniqueness. There is nothing alive like it today. One of the most interesting aspects of palaeontology is seeing how evolutionary pressures led to different solutions for the same problems, and in doing so created a multitude of beautiful forms. This amazing animal used to live on Earth. It's not from another planet, it's not from an artist's imagination—it slept under the same moon we do every night.

Rebecca Dart imagines what Horridus might have looked like, buried in the Hell Creek Formation.

> Life has been almost snuffed out by multiple mass extinction events, yet here we are.
>
> Rebecca Dart

Rebecca Dart's rendition of the Cretaceous forest in which Horridus lived.

The humerus is the upper arm bone. Although we think of *Triceratops* as having four legs, the forelimbs are classified as arms, made up of the humerus, ulna, radius and carpal bones in the hands. The hind legs contain the femur, tibia, fibula and tarsals.

For us humans it's important to understand that we are but one—currently short—chapter in the book of life on this planet. Life has been here for an unfathomable amount of time. It has evolved into all kinds of forms, from worms that exist in extreme conditions on the ocean floor to an albatross that lives on the wing for a month. Life has been almost snuffed out by multiple mass extinction events, yet here we are. An exhibition like this informs us about times long gone, but also shows the path travelled to where we are today, and where we might be going next.

I can't wait to see how future paleoartists will be inspired by the fossil and the art the team created for the exhibit. It's thrilling to think about being even the tiniest spark that ignites a creative passion in someone else. That's true artistic legacy.

Some of the welding involved delicate, intricate work.

Mark Radburn

Love on the Spectrum (ABC + Netflix)

At home, my pride and joy is the huge model of *Triceratops* that I bought at the National Dinosaur Museum in Canberra. It sits in my room, waiting to welcome guests. I love my *Triceratops* so much that I treat it as if it were alive. Every time I look my *Triceratops* in the eyes, I feel transported to its world, the late Cretaceous.

An exhibition about the natural world is a gateway to the past. It teaches us about who we are, where we come from and how we fit into the chain of life. The past holds all the answers. I hope that people young and old will be inspired by this exhibition to learn more. What we learn now will be carried into the next generation, and who knows what discoveries kids will make a hundred years from now?

Triceratops is such an iconic animal. It was destined for fame. We think we know a lot about *Triceratops*, but we've only scratched the surface. There is still so much to learn about the ceratopsian family. I hope more specimens will be discovered. I wish John Bell Hatcher, Othniel Charles Marsh and Edward Drinker Cope were still alive to see how far *Triceratops* has come and how much we have learned about this fabulous animal. I bet they would be very proud. *Triceratops* will never lose its awe-inspiring appeal and I hope it continues to fascinate future generations and make us want to learn more about our friend Mr. Three Horn. *Triceratops* is the awesome dino with the Tricera-Chops!

This specimen of *Triceratops* has dino-sized potential to give us an incredible amount of scientific data. Do you think we know everything about *Triceratops*? Wrong! Horridus is going to teach us more about *Triceratops* than ever before.

The bones of one individual can tell us about the palaeoenvironment(s) through isotopic sampling of different layers. Collectively, the bones of a single horizon reveal the biodiversity that existed at one time, and possibly their interactions. A series of horizons will give you change in biodiversity over time. It might be that bones grow like tree rings—and tree rings show hard winters and long summers of growth—so maybe bone can recall the same things?

I hope that kids will fall in love with *Triceratops* and all dinosaurs just like I did, and will want to learn about palaeontology. When people see Horridus, they will remember the encounter for the rest of their lives. The impact dinosaurs have had on my life is hard to fathom. I have learned so much and yet I want to know more. Dinosaurs are my family. I hope to meet other like-minded people that have a burning passion for dinosaurs and prehistoric life. I'm forever grateful to all the palaeontologists, past and present, who made me who I am.

The first left dorsal rib is carefully prepared for installation.

NMV P 25687

This block of three fused cervical vertebrae (known as the syncervical) forms the front of *Triceratops*'s neck, which had to support one of the largest heads in history.

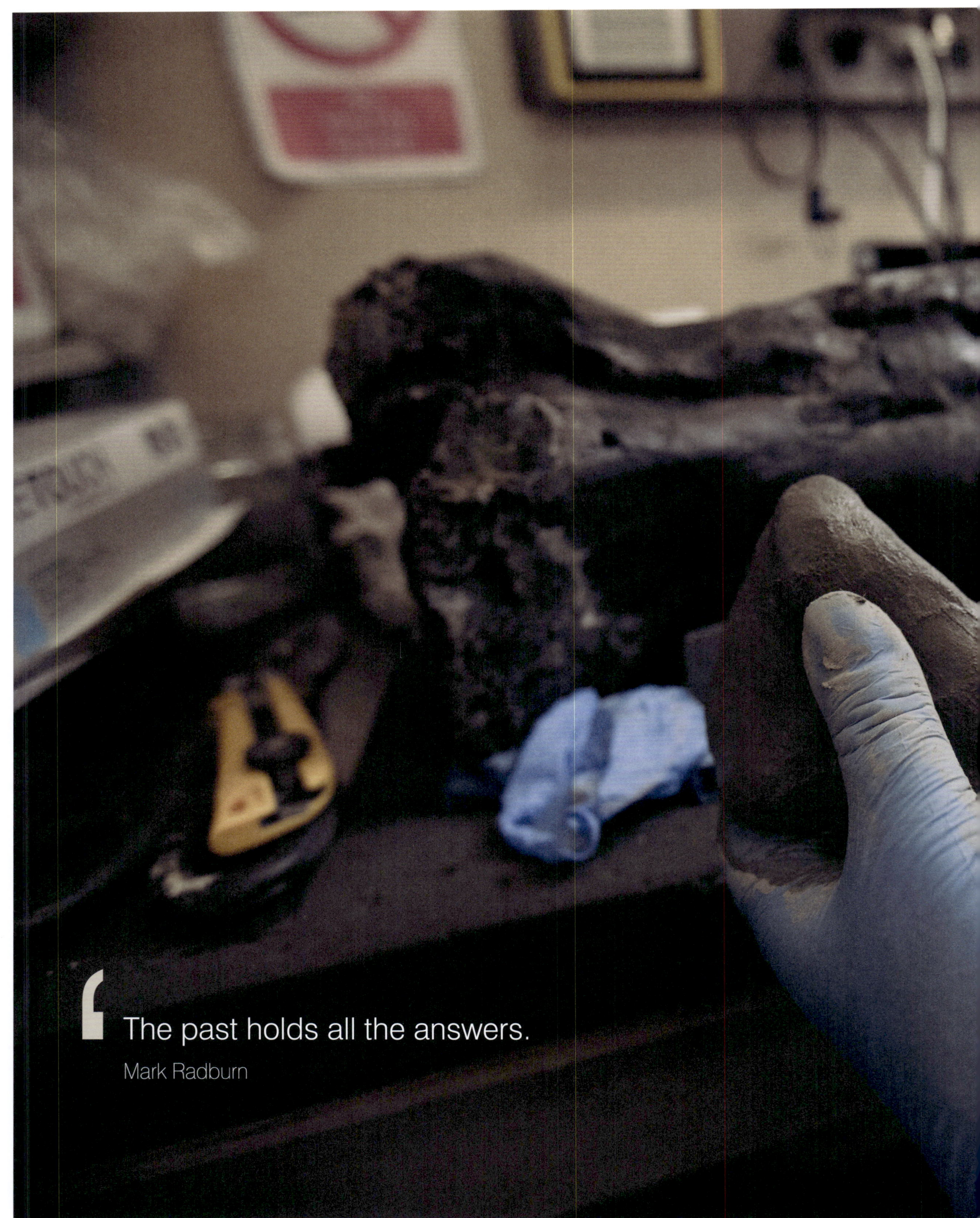

> The past holds all the answers.
>
> Mark Radburn

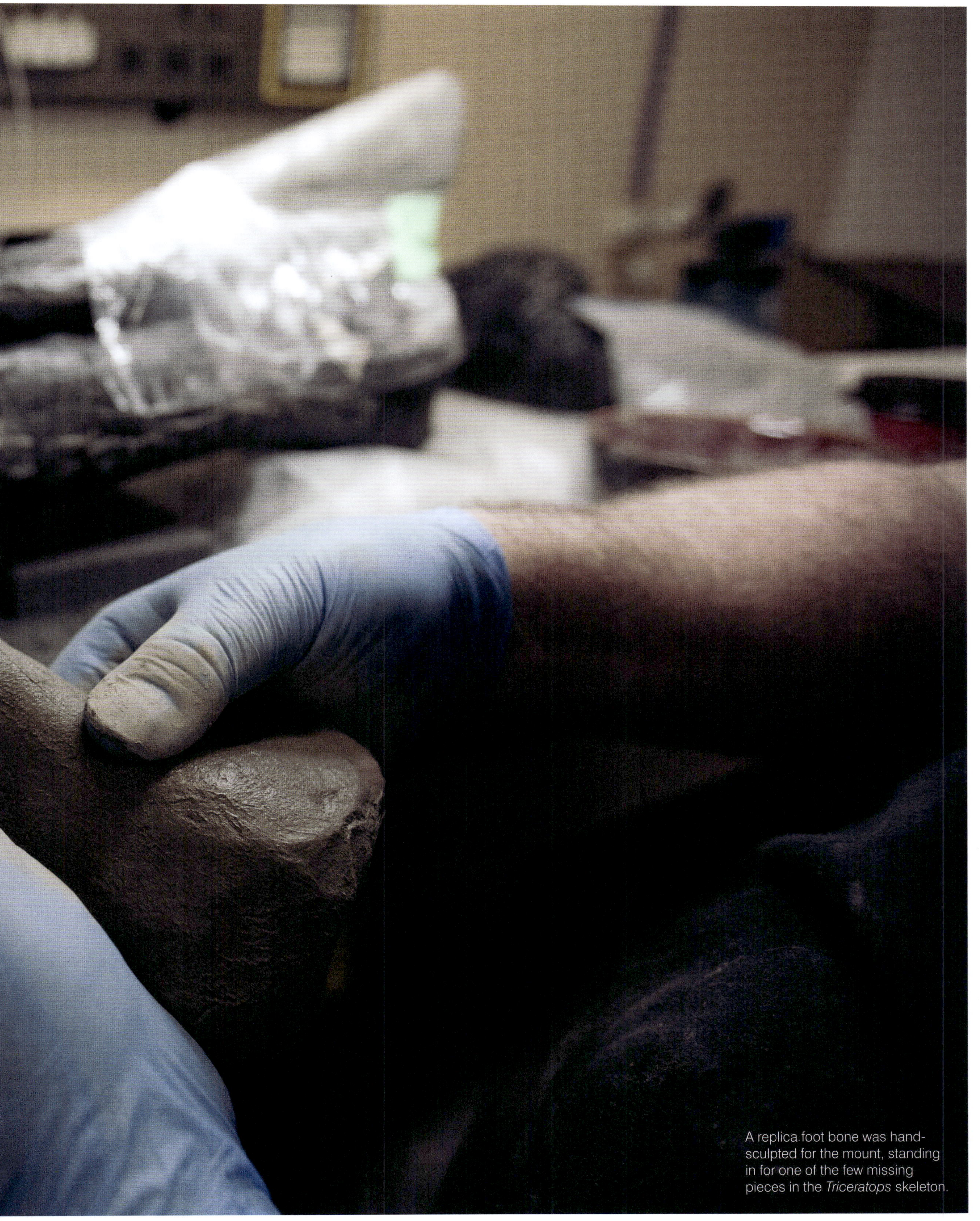

A replica foot bone was hand-sculpted for the mount, standing in for one of the few missing pieces in the *Triceratops* skeleton.

Tony Birch

Author + Activist

The dramatic change for me in understanding dinosaurs came through being a parent. The way they're regarded in popular culture as monsters is extremely problematic. It fails to recognise that dinosaurs played an integral part in the development of the planet, but this can be addressed through science, museums and education.

Strangely, by perceiving *Triceratops* as either a menace or engaged in battle, it has a correlation with the idea of contemporary Western human society controlling nature. Ninety-nine and a half percent of bogong moths have been wiped out, so we have very real contemporary stories of the eradication of species, and we're doing nothing about it. Can a story about dinosaur extinction give us a wake-up call?

Sixty-seven million years is almost beyond comprehension. Beginning with a story of a specific animal like Horridus helps people invest in history on an individual level. That is relevant to how we think about climate. When we consider the difficulties we face, for some people it's beyond their comprehension to know how to deal with it. One of the best ways to talk about climate change is to individualise and localise the story. In the same way, when talking about the reality of species extinction, you've got to personalise the narrative.

Where did extinct species fit? What other animals were dependent on them, and what was the interrelationship? This moves away from the notion of dinosaurs being evil monsters and accepting that they were integral to the ecology. What happens when species disappear? We never know until it's too late. The knock-on effects are inevitably greater than the hypotheticals, but it's always devastating. Look at the Tasmanian tiger. It was hunted down to near extinction, then you had the last few examples pathetically living out their lives in cages. Even though we all know what happened, it hasn't stopped us from making the same mistakes over and over again. We have to question our capacity to really change.

Debate continues regarding the function of *Triceratops*' horns. They may have been used for defence, or for fighting each other to attract a mate.

The left eye socket, warped from its natural circular shape by millions of years of heat and pressure to be slightly ovoid.

The ilium is the upper portion of the hip bone and pelvis. It is one of the longest bones in the skeleton.

What happens when species disappear? We never know until it's too late.

Tony Birch

Dr Susannah C. R. Maidment

Senior Researcher
Department of Earth Sciences
The Natural History Museum, London City

As a child I loved dinosaurs. I had an inflatable *Stegosaurus* and a *Stegosaurus* moneybox. When I was about seven or eight years old, my grandfather asked me what I wanted to be when I grew up. I said I wanted to be a scientist, but at that point I hadn't realised there were different sorts of scientist. My grandfather suggested I should be a dinosaur scientist because I loved dinosaurs. So I did.

Dinosaurs like *Triceratops* are so different from anything alive today, and often quite similar to depictions of monsters and dragons. They capture the imagination of children like no other creature, extinct or extant. Because of this, they're a great vehicle for educating children about the natural world, evolution and extinction.

Having a really complete skeleton to work with allows us to move beyond the bones and understand aspects of an animal's life that we can't get when we are working with fragmentary skeletons. For example, we can create a digital model of the bones and calculate body mass. An animal's body mass can tell us all sorts of things about the way it lived, from the size of its home range to its physiology. We can also use such models to better understand how dinosaurs moved and what maximum speed they could attain, which can tell us about how *Triceratops* might have escaped predation and whether it could run away from *T. rex*, or whether it had to turn and fight.

If we are to understand range-size change and extinction risk for large terrestrial vertebrates as our Earth warms, we need empirical data which can only be obtained by looking at how such vertebrates have responded to climate change in the past. This is why palaeontology has never been more important than today. By examining how dinosaur diversity changed in response to changing environmental conditions, we can seek to understand the causative mechanisms behind extinction.

I work on dinosaur locomotion. I have a theory (and it is only that—I haven't been able to test it yet) that large herbivorous dinosaurs couldn't run—they could only move slowly, and this is why they developed frills, horns, spikes and crests. They needed them to frighten off predators, or to fight back, because they couldn't run away. *Triceratops* is often depicted as being able to gallop, a bit like a rhino. I would love to get my hands on a digital model of the *Triceratops'* legs to test how fast it could move.

The left femur (long) and right femur (short), attached by brackets to their armature before being fitted to the mount. The femurs of Horridus have become distorted over the last 67 million years underground, to be significantly different in length.

A close-up of the *Triceratops* mandible illustrates its ability to cut through vegetation.

> Dinosaurs like *Triceratops* are so different from anything alive today, and quite similar to depictions of monsters and dragons. They capture the imagination of children like no other creature, extinct or extant.
>
> Dr Susannah C. R. Maidment

The spectacle of *Triceratops* quickly becomes apparent upon viewing the specimen in complete articulation.

Blair Holtner

Lead Fossil Restoration Technician
Dino Lab

I'm a cabinet maker by trade. It's incredible the crossover between that and what I do now. There's a lot of similarities between wood and fossils, in how you work on them. Woodworking gave me a lot of insight into techniques and how much force you have to put on a matrix to remove it from bone. I try to teach people how to see bone. You look at colour, texture and grain, just like with wood. I can look at something and see how it should be in three dimensions. Even if I only have a little bit of bone, I can visualise the whole animal quite easily.

Fragile bones required careful restoration.

A variety of delicate brushes were used to clean the specimen.

This caudal vertebra formed part of *Triceratops*'s tail. This is the very last caudal vertebra—caudal number 36, the tip of the end of its tail. This bone has never been found for *Triceratops* before, and helps us see for the first time how long their tails were.

> Even if I only have a little bit of bone,
> I can visualise the whole animal quite easily.
>
> Blair Holtner

A close inspection of the left humerus reveals the skill required in correctly piecing together fragments of fossilised bone.

Shaun Tan

Author + Artist

'How did it die?' asks my daughter.

We are standing before the display of skeletons at Melbourne Museum's Dinosaur Walk. She is five years old, an age when pointed questions about death, about beginnings and endings and bones, naturally rise into consciousness. As a well-read dinosaur nerd from way back, I'm more than happy to respond with a dutiful Dad-splanation, no less rote for being thoughtful: 'Well, most scientists believe that a huge rock from space crashed into the Earth and threw up a whole lot of fire and dust and stuff which blocked out the sun and–'

'No, no, no,' she interrupts. 'How did this one die?'

I suddenly realise that she's talking about a specific dinosaur. This specific dinosaur, the one we are looking at, a *Tarbosaurus* from somewhere in Late Cretaceous Asia. I also realise it may well be the first time I've ever thought about this myself when looking at a dinosaur skeleton. Ever since my palaeontology-obsessed older brother turned his childhood bedroom into a fossil museum, peppered with tiny handwritten labels and diagrams, I've been accustomed to seeing all petrified shells and bones as representative of a species, genus, family, class, order, phylum, kingdom … just part of that broader, anonymous narrative of natural history. Each greyish-brown ammonite or trilobite a specimen denoting a group, an illustration in stone, a hard condensation of time, an example.

I've never really stopped to think about the individual animals. Those very particular, very extinct lives: short, fleeting and vital. My daughter's simple question, perfect as so many children's questions are, opens a world of imagined narrative previously locked by generic conclusions and textbook explanations. How, for instance, did this particular *Tarbosaurus*, its skeleton now striding across a weatherless display room under halogen lights, actually live and die? What was its life story? What things did it see and hear, what did it fight and kill, what did it love? Did it have brothers and sisters? A mate, eggs, children? And what, my daughter is asking so matter-of-factly, happened at the end? What brought down this startling animal before us, elegant legs and spine raised from the grave, teeth still looking ready to strike some 70 million years later? It's as if we are attending the world's longest wake—my daughter would leave a bunch of flowers by its talons if she'd thought to bring one.

As with all funerary things, I think about my own mortality and shared continuum, something that I've been pondering more than usual since becoming a parent. The way creatures are born, reproduce and die, the unpredictable way that life goes on. As my daughter absorbs the new world around her, I'm considering my own ageing, and how generational memory is passed on. How it wanes and changes, rarely preserved intact with each passing of life's baton, and how things are often forgotten altogether. Everything becomes subsumed by water, bacteria, mud and ash one way or another. Everything except those most extraordinary of geological treasures: the fossilised remains of certain animals, dying in a certain way and leaving for posterity the most wonderful album of natural history—one we can see and touch. It seems miraculous that anything like this should remain for so long, and appeal so vividly to our simian imagination, even as we struggle to comprehend the chasm of time between us.

When we look at something like the skeleton of a dinosaur, that chasm collapses in an instant. We see in its ribs and eye sockets and joints and jaw a fellow creature. We can think about its passing—as my daughter does—as if it's yesterday, and our own passing tomorrow, and all the other individual threads in that massive, twisting yarn of DNA that has endured so many catastrophic events in the Earth's history. As we leave the museum, the pigeons—those plucky winners of nature's dice roll—strut before us on little reptilian claws. And when they stretch their wings and take to the sky, they do so with such ancient aplomb, as if things have always been this way, and always will.

The lower jaw of *Triceratops*, shot from above.

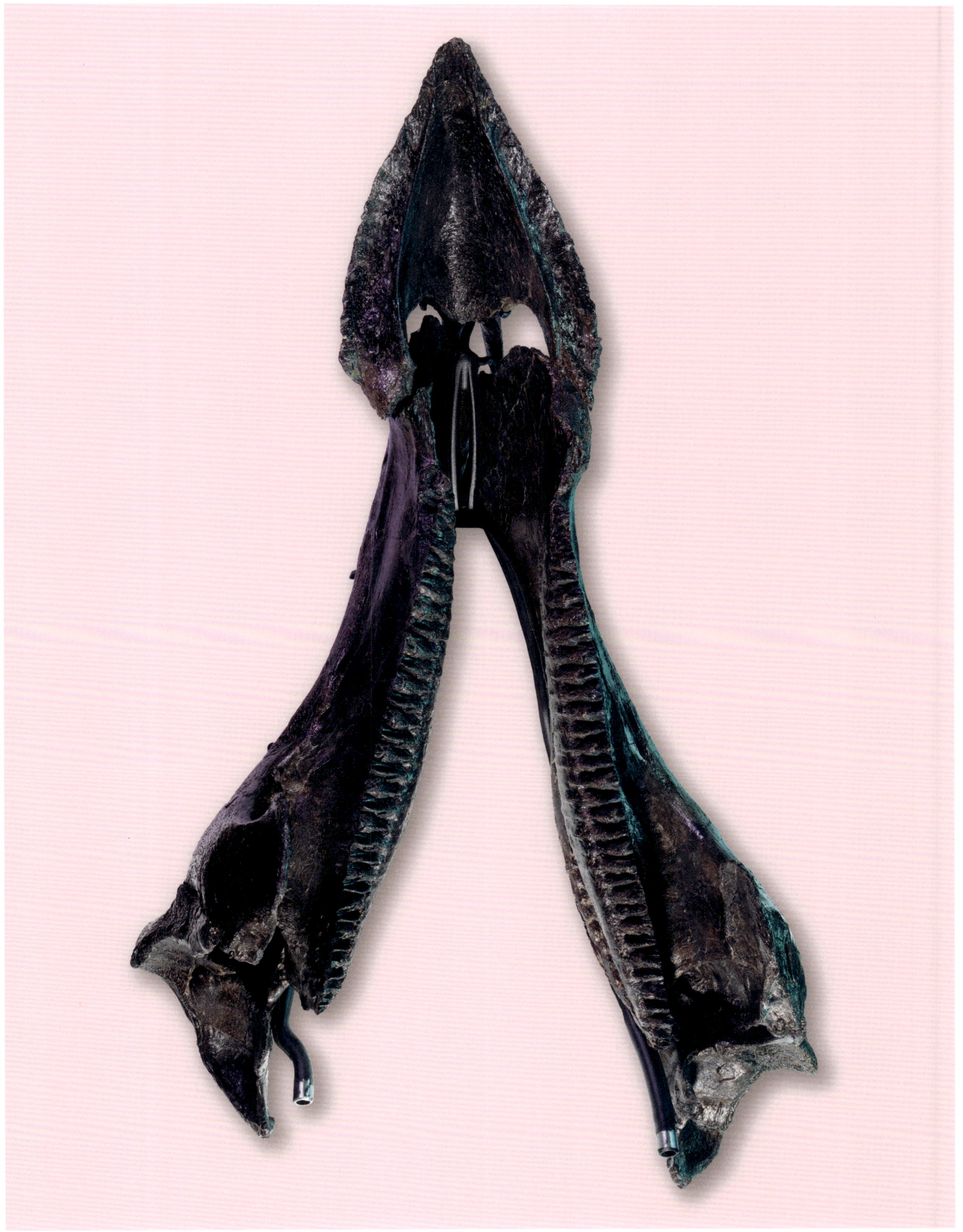

TAN

'Mother and child', pastel and charcoal on paper, 2021, 70x50cm

How did this one die? That is, how did this one live? As I consider the bones of *Triceratops*, this query again transports me straight back to my own childhood fascination with this species, an animal which probably set me on the road to being an artist more than any classroom or gallery. I remember being at home sick with measles, staring at a large painted poster from the Western Australian Museum of two *Triceratops* battling it out in an ancient, sunlit canyon. I've never been able to locate this image since, but that doesn't matter because it's so clearly etched in my mind, if anything enhanced by feverish delirium. What I recall most about the painting is a magical feeling of presence, like the momentary snapshot of a forgotten time, right down to the shadows drawn out from the heaving feet—hooves?—of these massive sunlit reptiles. 67-million-year-old shadows! A world lost in time, a place of vanished plants and animals and bugs, once illuminated by the same sun that shone every morning through my bedroom window. My planet was their planet.

Those *Triceratops*, or at least something very like this artist's dramatic impression of them, were real things. That single thought—the reality of an imaginary creature, laid bare in bones and footprints that set an artist's mind in transit from one reality to another—has never lost its hard grip on my imagination. I think it's the root of all painting and drawing, and science too: an ontological deep dive, a question of what reality is. Back then, having neither the ability or desire to articulate such big ideas, I simply wanted to draw those *Triceratops*, which I did, sitting in bed with the measles, trying hard to copy every flexing scale and *Pteranodon* wheeling above chalky cretaceous cliffs.

The articulated left manus (front foot) of Horridus, packed in a plaster jacket for transport.

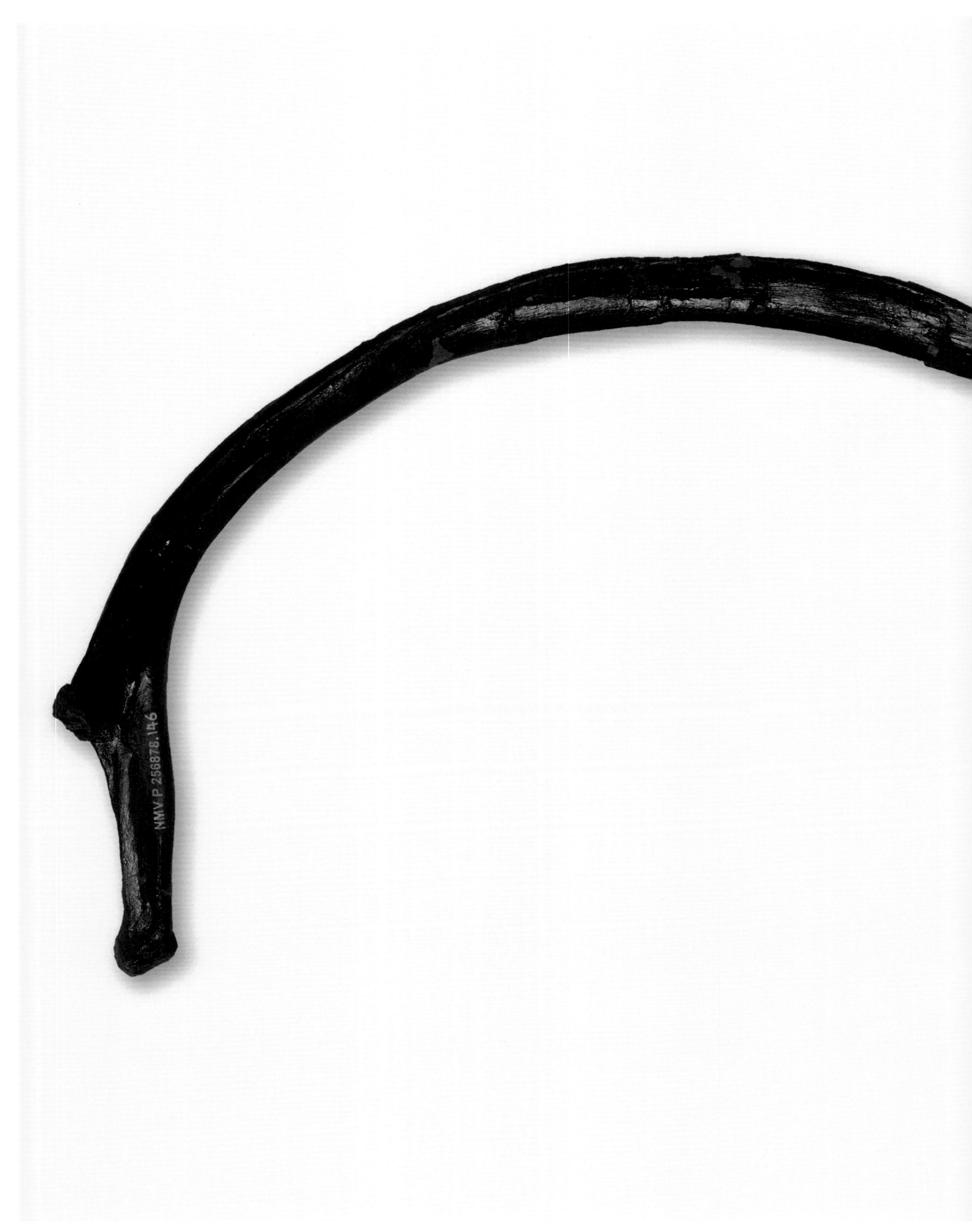
NMV P 256878.146

Mitch Mahoney

Boon Wurrung Artist + Cultural Educator

The museum tells stories of old things—tells stories about the past. It's here to conserve and keep those stories alive. From a First People's perspective, we're continually trying to research those stories and build upon them so that we know more about the past, but at the same time shifting that focus to how we relate to these stories today. How do we tell the story of today, and get people interested in caring about what's happening now? Then preserve that information, so that a couple of hundred years from now the museum of the future has a reliable, accurate story of what today is, what the items in the collection mean and the movements that we've gone through in our lifetime—socially and politically, but also in our natural world and how that's changing.

I see *Triceratops* as a lure, drawing people into the museum where they then wind up looking at everything else. But then I'm a fisherman.

For First Nations peoples, we don't think we own the natural world. We belong to the rivers and oceans and have a responsibility to care for them.

Triceratops would hate being alive today. They'd have trouble breathing the oxygen that we have. It's such a hostile environment. I feel bad for elephants living in our world, let alone some prehistoric dinosaur. Maybe *T. rex* would be all right, or raptors. At least they could deal with people by eating them.

A fully intact dorsal rib.

The steel armature was meticulously crafted to support the bones without causing wear or damage.

> I see *Triceratops* as a lure, drawing people into the museum and then they wind up looking at everything else. But then I'm a fisherman.
>
> Mitch Mahoney

> For First Nations peoples, we don't think we own the natural world. We belong to the rivers and oceans and have a responsibility to care for them.
>
> Mitch Mahoney

When inspected closely, the mandibular teeth of *Triceratops* resemble a prehistoric landscape.

Thomas H. Rich

Senior Curator Vertebrate Fossils
Museums Victoria

Whether they had horns or not, the ceratopsian 'horn-faced' dinosaurs have a skull structure that makes them distinctive not only from other dinosaurs, but all other vertebrates as well. In addition to the horns, the lower arm bone (called the ulna or elbow bone), is so unexpectedly distinctive that it has provided clear evidence that these very ceratopsians were living in Australia 130 million years ago. Prior to that discovery, the ceratopsians were known almost exclusively in the Northern Hemisphere.

When the Victorian dinosaur ulna was first found by Mike Cleeland at the base of the Arch near Kilcunda, my first guess was that it was some kind of carnivorous dinosaur or theropod. This was because it was a short, stumpy bone, which is so characteristic of the forearms of *T. rex*. Many theropods have such reduced forelimbs. But I was wrong, as further information clarified.

I visited the Royal Tyrell Museum of Palaeontology in Drumheller, Alberta, Canada, where I was able to examine the ulnae of a number of theropods. None of these were even close in shape to the ulna found in Victoria. Then I walked past a skeleton of the ceratopsian dinosaur *Centrosaurus*. I noticed that its ulna was very similar in shape—only about twice the length—of the unidentified Victorian ulna. Knowing that much smaller ceratopsians than *Centrosaurus* existed, I asked staff at the museum if they had any of these other, smaller ones in their collection. The answer was that a skeleton of one of these smaller ceratopsians, *Leptoceratops*, was at that very moment being moulded and cast in downtown Drumheller. *Leptoceratops* had been collected not far from the Tyrell Museum in rocks about 60 million years younger than those from which the Victorian ulna came.

When the Australian and Canadian ulnae were laid side by side, a gasp went up from all concerned. If the two fossils had come from the same hole in the ground—instead of having been found 14,000 kilometres apart and from rocks differing by 60 million years in age—you might think they belonged to the same species.

The Victorian ulna was not only unique, but so close to a ceratopsian ulna that it had to be one. Eventually, this intriguing fossil became the type specimen for a brand-new genus and species of dinosaur, *Serendipaceratops arthurcclarkei*, representing a group previously unknown on the Australian continent—the ceratopsians. The dinosaur was named in honour of my favourite science fiction author and good friend Arthur C. Clarke, who said, 'Now I have a dinosaur and asteroid named after me, what else is there to live for?'

130 million years ago, when *Serendipaceratops* lived, Australia was far from North America and Asia where most ceratopsian fossils are known. How interchange could have occurred between Australia and those continents, even with Europe included as another landmass with a ceratopsian record (albeit a meagre one), leaves the identification of the Australian specimen a topic rife for speculation. Australia was more isolated from those three land masses then than it is now. It is perhaps a case of serendipity—the facility for making fortuitous discoveries by accident.

Do these hints suggest that ceratopsians flourished more widely than their fossil record of today suggests? The notorious incompleteness of the fossil record is one of the challenges that keeps palaeontologists scouring the global fields today. It is always to be kept in mind and challenges many to 'keep looking'.

The left (long) and right (short) femurs, removed from their supportive armature.

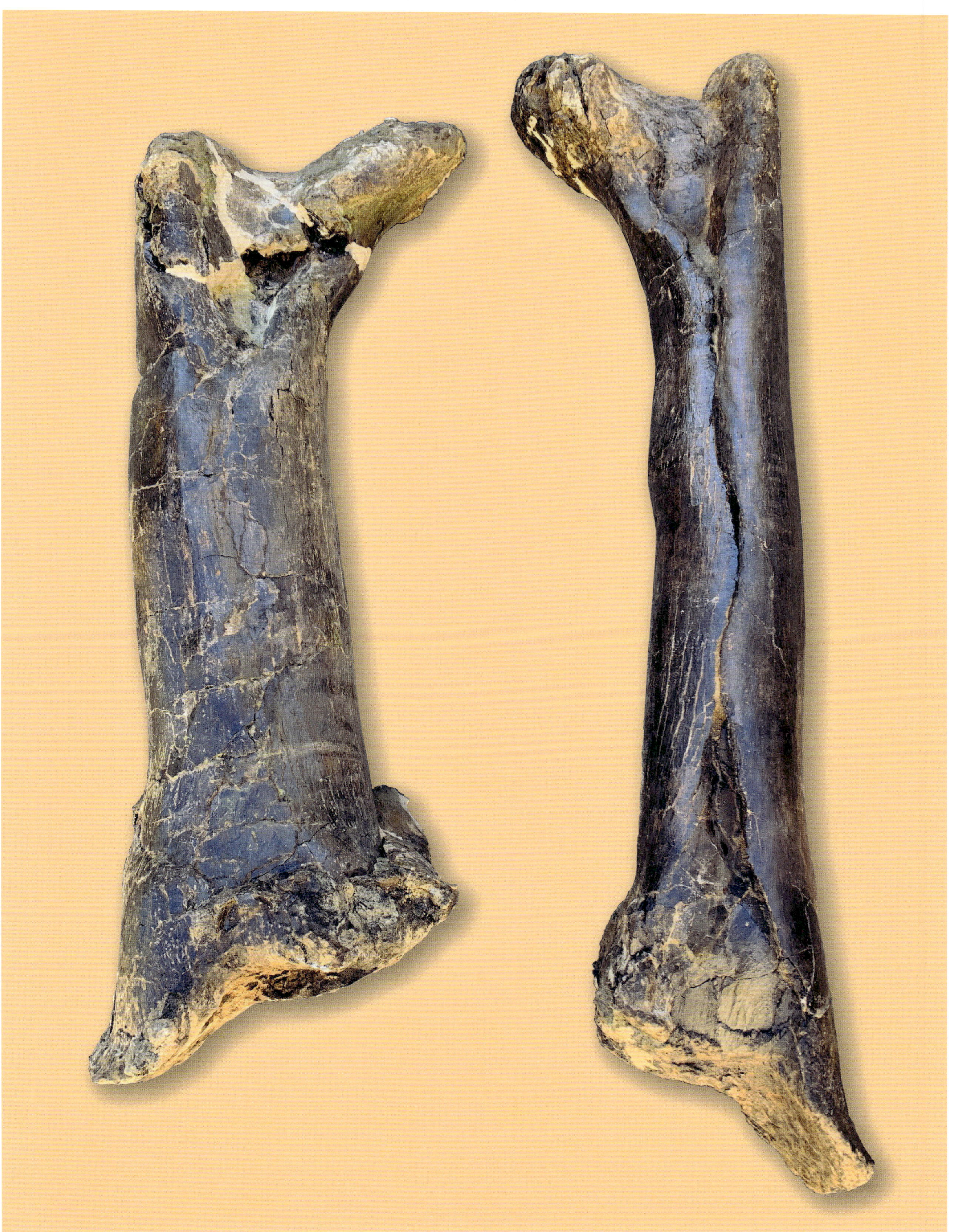

One of the major challenges in palaeontology is in correctly piecing together disparate bone fragments.

Extraction of the fossil from its stone matrix was a methodical and time-consuming process.

Shannon Martinez

Owner + Chef
Smith & Daughters/Smith & Deli

CRETACEOUS EMPANADAS DE PINO

EMPANADA DOUGH

3 cups plain flour

1 1/2 tsp salt

1/4 tsp baking powder

4 tbsp chilled vegan shortening, cut into small pieces

4 tbsp chilled vegan butter, cut into small pieces

3/4 cup cold water, or more as needed

FILLING

4 tbsp of butter

1 small brown onion, diced

2 garlic cloves, minced

1 tsp ground cumin

1 tsp sweet paprika

1/2 tsp chilli powder, or to taste

Salt and black pepper to taste

4 tomatoes, peeled, seeds removed and finely diced.

2 cups tinned heart of palm, finely diced

1/2 cup pitted black olives, roughly chopped.

Handful chopped parsley

Soy milk or vegetable oil for brushing

METHOD

First make the dough.

Mix the dry ingredients in either a regular bowl or the bowl of a food processor. If you have a food processor, add all the dry ingredients along with the shortening and butter and pulse until the mixture is the consistency of clumpy sand. If you don't have a food processor, use a pastry cutter or your fingers and rub the butter and shortening into the dry ingredients until the same consistency is achieved. Drizzle in your water while mixing. Only use enough water for the dough to hold together in a ball. Divide your dough into two parts and flatten into discs, then wrap and put in the fridge while you make the filling.

To make the filling, heat a large frying pan over medium-high heat and fry the onion along with a pinch of salt in butter until translucent. Throw in the garlic and spices and cook out for a minute, then add the tomatoes and cook until they start to break up. Add the chopped heart of palm and olives, along with ½ cup water and cook for about 15 minutes or until the liquid has almost completely evaporated. Remove from heat and stir through the chopped parsley. Cool to room temperature before filling the empanadas.

TO BUILD

Preheat oven to 180 degrees.

Cut the dough into approximately 20 pieces in total and roll into balls. You are welcome to make whatever size empanadas you like. Allow the balls of dough to rest for at least 5 minutes, then roll out into disks approximately ½ cm thick. Add 1 tbsp filling to the centre of the disk and run a wet finger along one side of the pastry. Fold over so the edges meet, then press to seal. Either fold up the sides or crimp with a fork. Brush the empanadas with a little soy milk or oil and bake for 20–30 minutes, or until golden brown.

A fossil leaf was found in sandstone
near the Horridus dig site in Montana.

Triceratops didn't 'grind' its food like a herbivorous mammal. The teeth worked more like serrated scissors to 'snip' it into small particles.

Each section of the specimen
was numbered and catalogued.

2B

Close-up of *Triceratops*'s mandibular teeth.

TRICERATOPS MENU

Grilled fig + pine nut cheese tart
(Angiosperm + Conifer)

Cassava stuffed with pickled fiddlehead and magnolia, steamed in banana leaf
(Fern + Magnoliaceae)

Hearts of palm empanada
(Palmae)

Rice pilaf with barberry and fermented wild onions
(Angiosperm)

Pandan and ginkgo nut sago
(Cycads, Ginkgophyte, Pandanaceae)

Sarsaparilla gin fizz
(Salicaceae + Conifer)

Ry Williams

Lead Welder/Dinosaur Levitation Expert
Dino Lab

My brother Nate and I started our careers welding high pressure pipelines for oil and gas. There's no handbook for creating armatures for cinosaurs. The ultimate level of craftsmanship lies in you not seeing what we do, because we want to show off the skeleton. Welding is a fine art. It's like crocheting tiny puddles of flowing metal.

The armature has to be safe for the bone and for the public. It also needs to be graceful. I want to see motion. We're trying to elevate the bones in the best way possible. I wound up having an intimate knowledge of every single bone from Horridus's body. I could see the injuries, the way the joints fit together. I could tell if they had arthritis or not. I could see where arteries and nerves went through the bones.

This *Triceratops* was a huge part of my life during the weirdest time in the world. I worked alone on this skull for a few months at the start of the pandemic. It was just me and Horridus's head. It's impossible to not create a backstory, because they're real—they were alive. This creature fought to survive, and that story comes to life when you see these bones. This is part of our history. *Triceratops* is the most iconic vegetable-eating creature that's ever lived. It battled against the most vicious carnivore on Earth, and it didn't back down. It's still here.

Ry Williams worked on both a small and very large scale.

SAP#11023135

> It was just me and the head. It's impossible to not create a backstory because they're real, they were alive.
>
> Ry Williams

The horns and skull were amongst the final pieces articulated by Dino Lab in Canada.

Dino Lab staff were amongst the first humans to ever handle this 67 million-year-old fossil.

The design of the armature mirrors the natural shape of the bones, holding them securely in place.

MVT.1.208.001
0

Dermot Henry

Head, Sciences Department
Museums Victoria

Over many years, I've been involved in acquiring amazing specimens for Museums Victoria. It's an enjoyable privilege to contribute to the development of the State's collections, which are a rich resource for interpreting the past and present, and for informing the future. The collections today reflect the work of many, over the past 168 years, who made and took opportunities to acquire and preserve specimens. The acquisition of this incredible *Triceratops* is an opportunity created by many. Horridus will be an inspirational specimen for generations to come. They will be examined and re-interpreted in ways we cannot imagine today, and they will continue to inform our knowledge of the evolution of life on earth.

Our palaeontologists are prehistoric detectives. It's CSI: Cretaceous—looking for clues, building a picture of how, when and where this *Triceratops* lived and died. There are samples from Horridus which may represent stomach contents. What did they enjoy as a last meal 67 million years ago? The rock surrounding the specimen contains other fossils—possibly even pollen, which build a picture of the environment where this *Triceratops* lived. The nature of the sediments, from around the bones, tells us about the landscape. Horridus fell in a river and was quickly buried. We can also find clues about their environment by looking at what plant and animal fossils are found in the rock formations around their tomb.

The phalanx is one of the bones in *Triceratops*'s front foot, or manus. This image shows two phalanges from the manus stuck together with matrix before they were separated (the smaller one on the right and larger on the left).

> The rock surrounding the specimen contains other fossils—possibly even pollen, which build a picture of the environment where this *Triceratops* lived.
>
> Dermot Henry

The specimen had to be carefully removed from its field jacket.

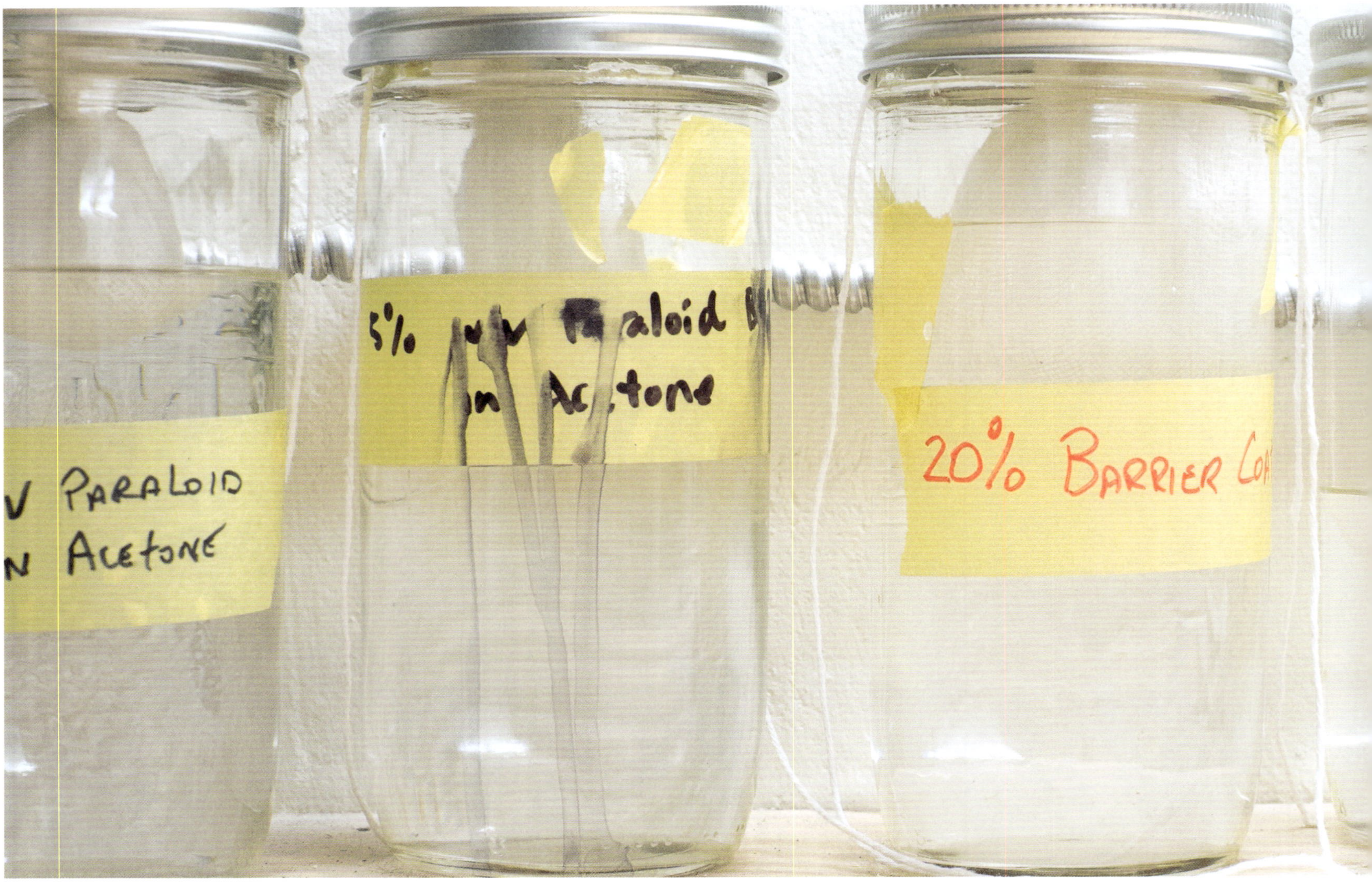
PARALOID
ACETONE
5%
Acetone
20% BARRIER

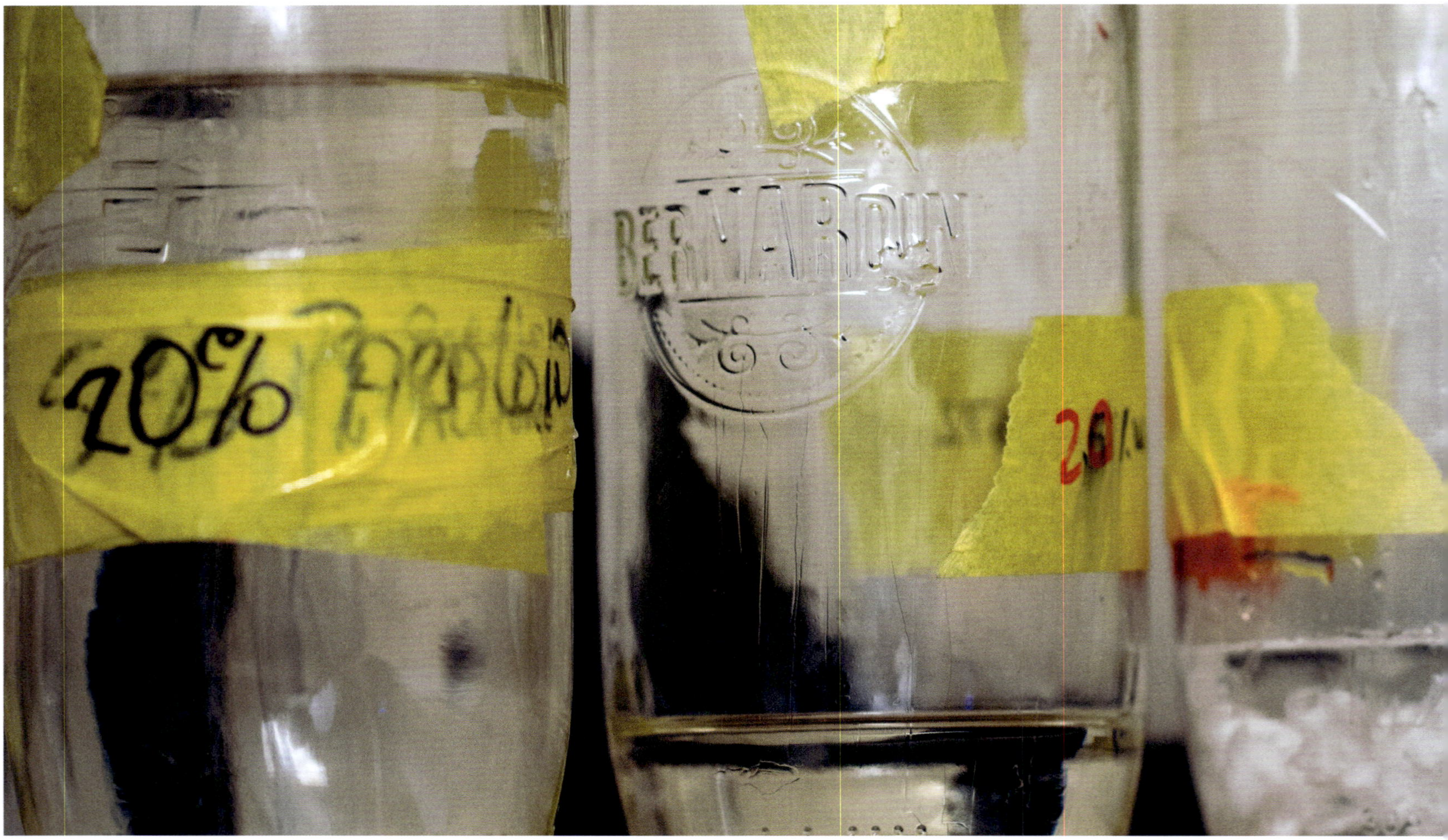
20%

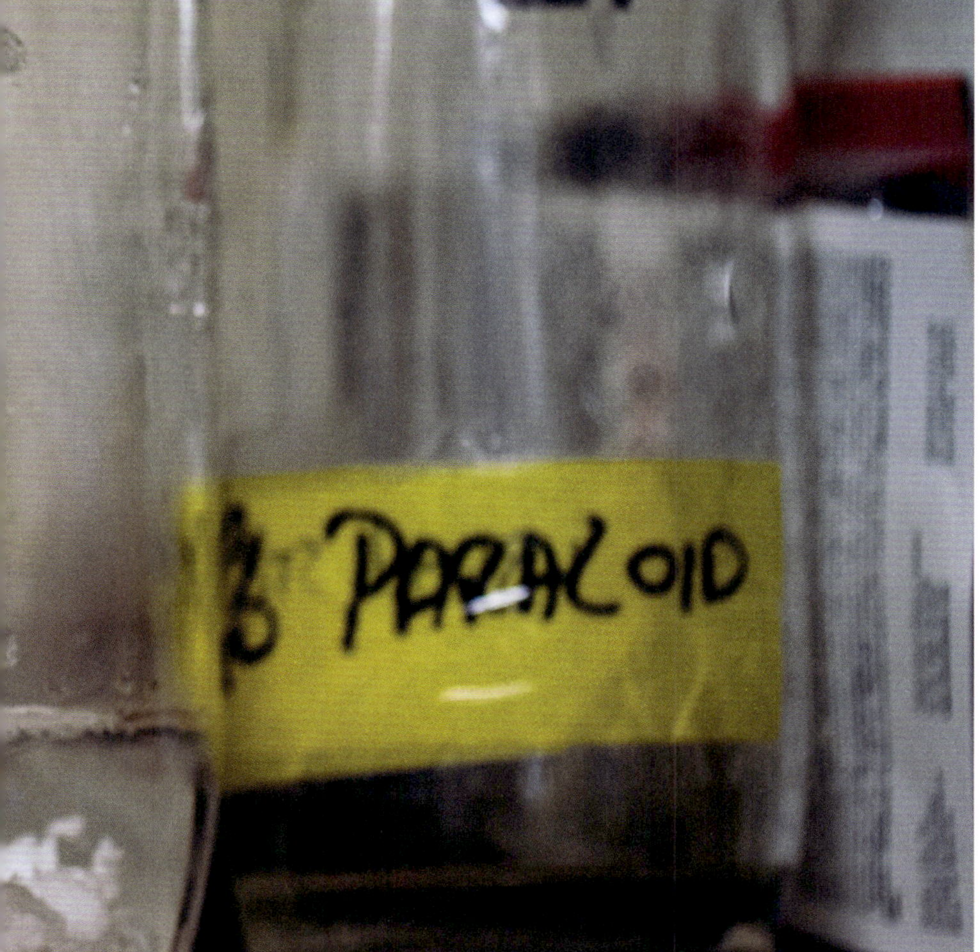

Fossils can help us understand the processes of evolution and extinction. Ninety-nine percent of all life that ever existed on the planet is extinct. We exist within a tiny window in time. Fossilisation is rare. If today, all life on Earth was suddenly wiped out not every species would be fossilised. If a visitor came to Earth in fifty million years to examine the fossil record, they would only get a snapshot of what exists on the planet today. And yet, life has filled every niche on the planet, from the deep oceans to the frozen reaches of Antarctica. There are life forms adapted to hot acidic environments and environments rich in arsenic and other toxic metals. In many ways, life is very robust—the question is how rapidly can it adjust to change?

When I look at dinosaurs—this one in particular—I wonder 'why does a beast like Horridus become extinct? Why did other species, such as crocodiles, prevail'? These thoughts link with the lottery of modern extinction in the Anthropocene. Since European settlement, Australia has a terrible record of species loss caused by rapid environmental change induced by human activities such as land clearing, destruction of habitat by the draining of wetlands and the damming of rivers, the introduction of non-endemic species and diseases, non-sustainable harvesting practices such as the overfishing of the oceans and through the effects of pollution. Species are disappearing every day, including species we didn't even know we had. Not enough people are reflecting on the cost of biodiversity loss.

We've become too blasé about the beauty of nature, yet tomorrow it might all be gone.

Jars of adhesive and barrier solutions were used to protect the fossil and prevent it from crumbling as it was handled.

Conservator Danielle Measday measures radiation in the *Triceratops* remains with a Geiger counter.

Sixty seven-million-year-old fossilised bones often look remarkably fresh, as if the *Triceratops* had only died recently.

Laura Jean McKay

Author

I was a child of the 80s and at our rural primary school in Sale, Victoria, dinosaurs were THE THING. The cool kids knew about them and the rest of us just listened in awe. There was one boy—the coolest—who knew the most about dinosaurs. We were all building up to a special trip to Melbourne Museum where a dinosaur expert was going to show us fossils. When the expert asked a question like, 'And when were dinosaurs alive?' everyone looked at the kid and he just froze, mumbling an answer about as useful as, 'The 1800s?' Things weren't the same after that.

This *Triceratops* exhibit shows us that extinction has always occurred, but not at the rate it is now. While species have lived and died out before, we are living in a time where extinction is occurring across species at a terrifying rate—and much of this is caused by humans. *Triceratops* offers the opportunity—in this overwhelming time—to stop and contemplate extinction. The species that have come before, the species recently extinct, and those that are critically endangered and can still be saved.

A *Triceratops* might find Melbourne hard and frightening. Looking for green, they might head straight out the doors of the museum and into Carlton Gardens. In lockdown times, this is what a lot of people have longed for too—some green open space. If *Triceratops* saw these people, maybe they'd ignore them—not realising that humans are the ones to watch out for. Instead, *Triceratops* might try to take on one of the trams or cars, seeing them as an obvious threat. After flipping a few cars, I hope *Triceratops* would find their way out to the highland forests, where it's a bit more like home.

The skull of Museum Victoria's *Triceratops* is 99% complete, an unprecedented find.

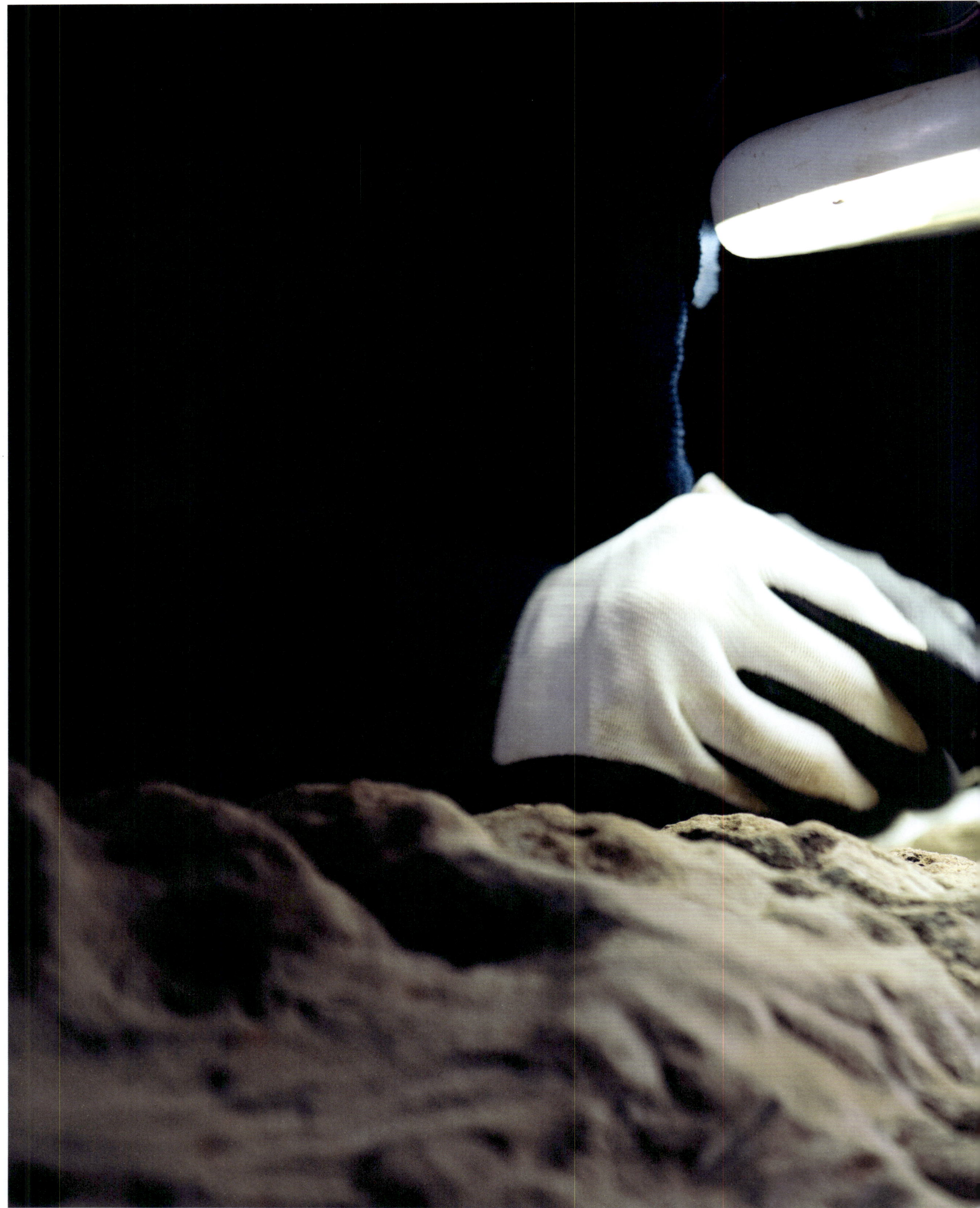

Restoration and conservation of fossilised remains takes enormous patience and care.

> *Triceratops* offers the opportunity—in this overwhelming time—to stop and contemplate extinction.
>
> Laura Jean McKay

Each rib had to be individually mounted and secured.

The vomer forms part of the palate, deep within the skull.

The predentary bone, which made up part of the lower beak, would have helped *Triceratops* gather food. The predentary is a unique feature of the group of beaked dinosaurs to which *Triceratops* belongs.

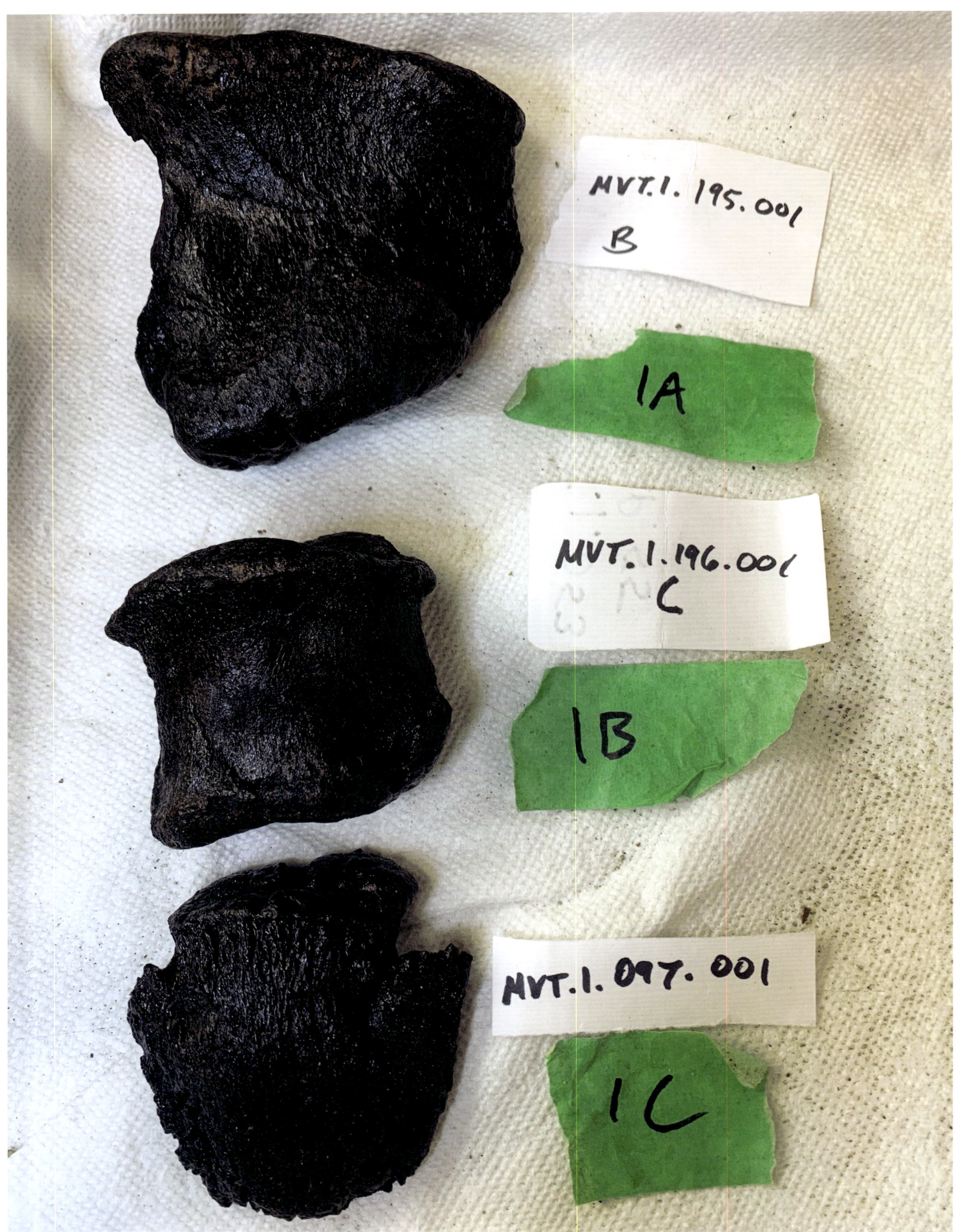
MVT.1.195.001
B
1A
MVT.1.196.001
C
1B
MVT.1.097.001
1C

Tim Ziegler

Vertebrate Palaeontology Collection Manager
Museums Victoria

We've joked about writing a book about this when everything is done and dusted. The working title is *How to Buy a* Triceratops *in 300,000 Easy Steps*.

It's about taking the holistic environment that the specimen is preserved in and really starting at the surface—at the end of your hammer, the tip of your pick. We'll work our way back through time, peeling back all the different layers of environmental influence from the geological record. We can look at how our specimen was fossilised, or use surrounding rocks to explore where and how it became buried, or look for traces on its bones to consider if it was attacked by predators or consumed by scavengers. What can we understand about the whole story of that scientific specimen by going all the way back to when it was alive? That's what taphonomy does. It's a forensic science. It's trying to piece back together the whole story from a partial record. You only have what you see in front of you, but we restore it to the three dimensions of the world and pack those layers of time back onto it.

Some people will look at a bone or tooth and see a treasure. They admire its physical features: the colour, the surface textures. I think that's interesting. Every environmental factor is written into that object's aesthetic. By applying different methods of analysis, you can draw information out of a static physical item back into a timeline of something that died tens of millions of years ago.

Witnessing this *Triceratops* is an opportunity to experience the sublime—that feeling of being much smaller than the world around you. Standing underneath a mountain ash in a rainforest in Victoria, seeing the biggest flowering plant in the world and feeling awed and shrunk down by that experience—that's what I feel working in palaeontology. On our human timescale, you're in contact with an infinite amount of time.

One of the benefits of working with such a scientifically complete specimen is that you can create a dynamic, realistic and authentic presentation of it in the exhibition gallery. This looks like a living animal. People will very quickly lay their own interpretation of character onto Horridus because the specimen is so authentic. I hope that people appreciate that yes, this is a fossil, but it's not so different from the remains of that animal when it was alive.

An interesting aspect of popular, well-known dinosaurs like *Triceratops* is that public interpretation is often so wrong—so dated. Since the middle of the 19th century when the first specimens were collected in North America, there was a desire to not only understand these specific things as bones or fossils but to get in touch with how the animal was in life—to speculate on what that life might have been like. If you do that from a skull, a femur, or four vertebrae in isolation, inevitably your imagined version is more speculation than interpretation. The step change in placing large, complete, well preserved and carefully documented specimens like this into the public collection gives the opportunity to overturn centuries of assumption about an icon of natural history.

For many kids, palaeontology and dinosaurs are the first places where they feel like they know something more than their parents. That's a step in their development, where they gather knowledge for themselves and gain authority. Palaeontology is tangible because you can put your hands on it in the ground. It's a gateway to scientific investigation. My path into this world came by recognising that I love being outside. I love interacting with such a complex, dynamic and beautiful thing as the surface of the planet.

One of the things that is a fantastic new outcome from this project is that we can reconstruct digitally what this animal looked like buried in that consolidated sand out in Montana. It's a unique example—an exceptional presentation. It goes against the preconception: the common experience of what fossils look like as being fragments of scattered bones. It blew me away when I saw mock-up versions of the specimen in situ using skeletal scans. This animal slumped down like an old dog on its haunches and went to sleep. It's like coming across the animal still living in that landscape. Being able to see it in that way is a result of comprehensive data collection from the outset, from the first inspection of the specimen in the field all the way through to when it goes on display.

The metacarpal and phalanx bones (phalanges) are part of the front feet of *Triceratops*. Pictured is the 'thumb' or first digit finger.

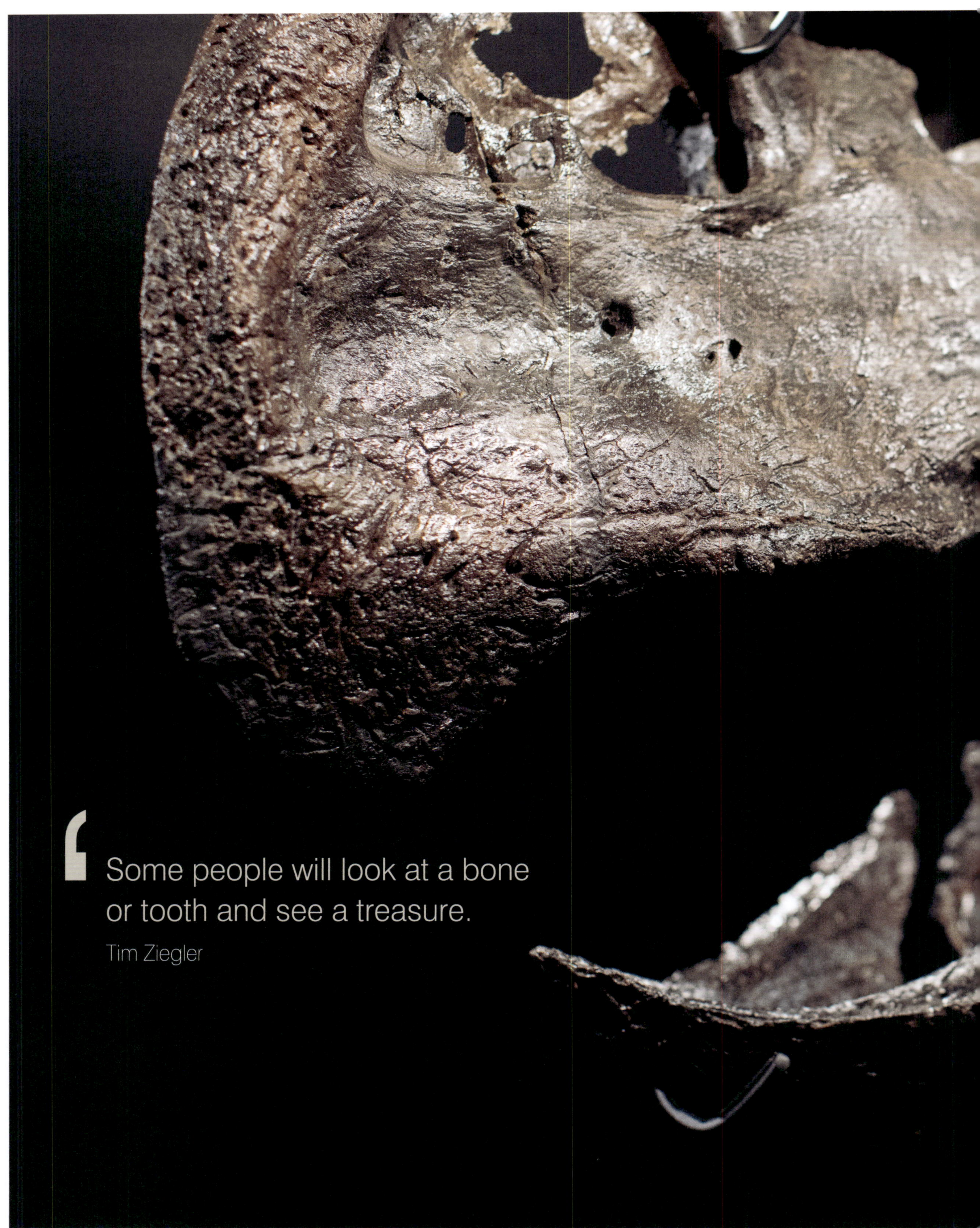

> Some people will look at a bone or tooth and see a treasure.
>
> Tim Ziegler

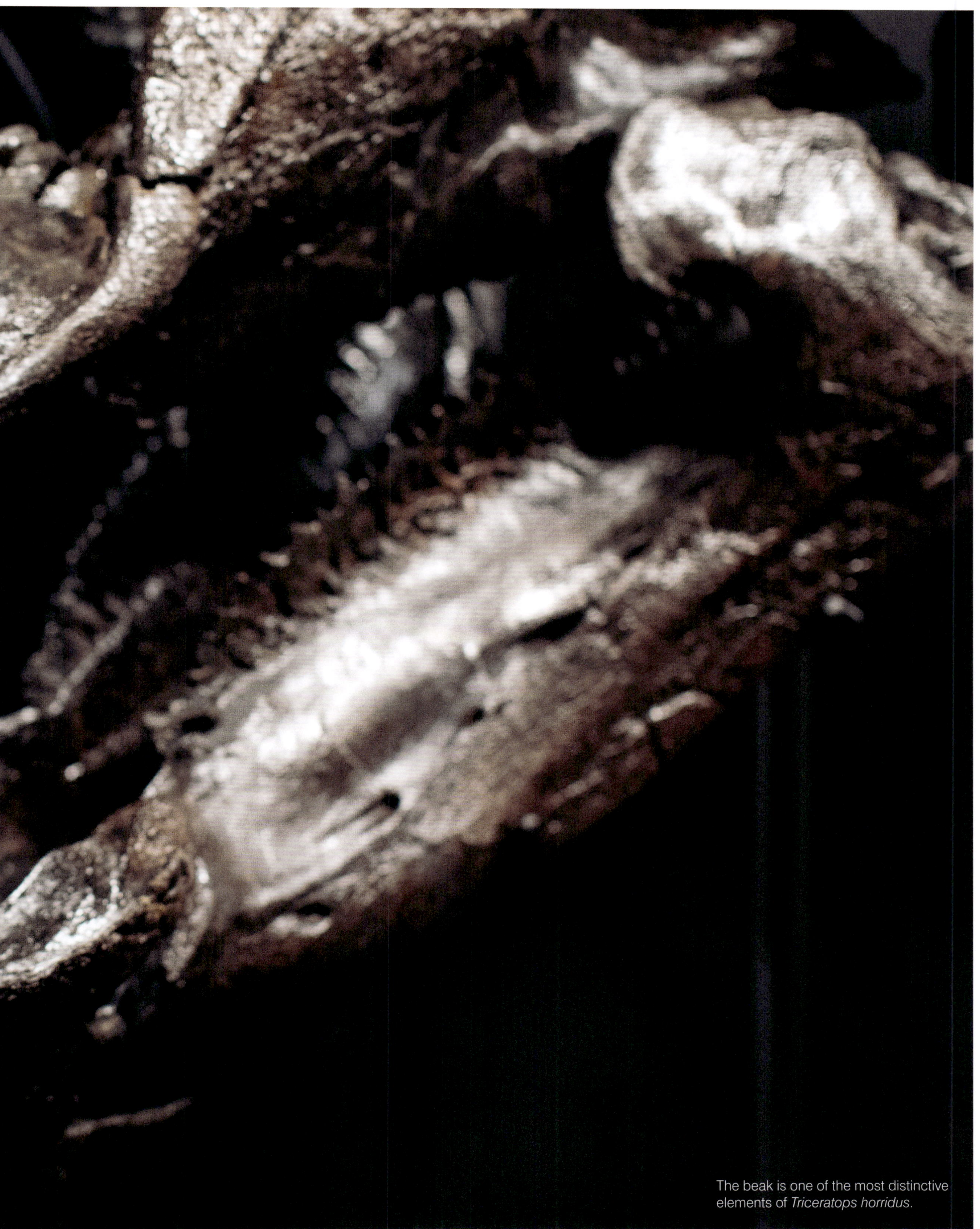

The beak is one of the most distinctive elements of *Triceratops horridus*.

SIEMENS
Healthineers
MONASH
University
VBIC
SCIENCE AND
INDUSTRY
ENDOWMENT
FUND
SOMATOM go.Up
CAT

We've created a complete electronic catalogue of the specimen. All the individual bones and fragments have unique identifiers. Every bone is labelled individually. If you peer very closely in the right spot, you might be lucky enough to spot one of these little strips that we've got hidden away. All the information we've gathered is part of the public archive record, which ultimately belongs to everyone.

Going to Monash Biomedical Imaging in Clayton to scan the *Triceratops* was a fun experience. The CT scanner is normally used for human patients. They've had antelope skulls and frozen wombats but never a *Triceratops*. In Melbourne Museum, we have an insatiable desire to analyse and gather data on specimens. There's no end to it. We'll drill down to the finest detail, so to see inside the bones—to see the internal structure preserved inside the skeleton—was another level of access that really brought the animal to life. You're seeing perforations for nerves and blood vessels, empty spaces for fatty deposits from when the animal was alive, all preserved internally. That will be accessible to visitors in the gallery experience. This is the first time any human has ever seen inside these bones. It's access to a whole other dimension of the natural world.

I'm constantly humbled by working with this part of natural heritage. I'm privileged to have my job, and I feel like I have a responsibility to impart as much of my experience to others as I can.

I'm not allowed to have a favourite bone. The humblest knuckle bone is just as important to the whole story as the three famous horns. Having said that, *Triceratops* doesn't have a chin. They have an extra bone that makes the lower beak, and the way the predentary bone slots in reminded me of a parrot. Placing that one bone in the steel armature made Horridus suddenly spring to life.

This animal lived and died, had its experiences, and now its remains are part of natural heritage. But for us as humans, we look at it and see something more. We put our own layers of meaning onto it. In the museum, we can guide what that understanding might be for someone. This exhibition is a chance for people to appreciate the wonder of the natural world and the depth of time. All the complexity and beauty and joy of biology and ecosystems around the world—from rainforests, the Great Barrier Reef, deserts, mountains, deep down in the Mariana Trench—life everywhere is a dynamic and amazing thing, but as a palaeontologist that living world only scratches the surface. Countless species have evolved and gone extinct. Environments have changed over time—sometimes gradually, sometimes rapidly. It illustrates the fragility of today. The only thing you can guarantee is change. This animal lived and died before *Triceratops* as a genus and *horridus* as a species went extinct. Its experience is within that span. The difference between that mass extinction and the sixth mass extinction going on right now is that we know it's happening. We have collected the data. It's undeniable that species are going extinct at a rate far above normal background process, and have been for years. The evidence overwhelmingly shows that this is a direct result of our activity on the planet. We're the influence. We're the new factor. We are the change in the world that is causing modern mass extinction. I hope that people can appreciate how fragile a seemingly robust and well-adapted system is to change and understand that we are as much a part of the natural world as anything else. We believe ourselves to be outside the natural world—not just atop it, not just controlling it, but somehow independent from it. The world is proving us wrong. And we would do well to pay attention.

When I die, bury me somewhere like a *Triceratops*: at the bottom of a river channel with a big layer of sand on top. Maybe I'll still be there in 67 million years and someone will dig me up. I hope they make a beautiful armature for me. They'll find me in my natural environment, with a hammer in my hand. Or maybe they'll mistake the hammer for part of my arm. Frankly, that would be even better.

All the effort that has gone into acquiring this specimen will be followed by work in maintaining, conserving, preserving, and making it useful and meaningful to people forever. It's a vast undertaking. But that's what museums do. We make the impossible real. It is so unlikely that we have this intact specimen after 67 million years, but everything in the natural world is unlikely. The odds of anything happening are astronomically low. I hope that adds value to people's perception of themselves and their perception of the world. Everything is so precious, just by virtue of being here.

Tim Ziegler at Monash Biomedical Imaging, loading fragments of *Triceratops* into a CT scanner.

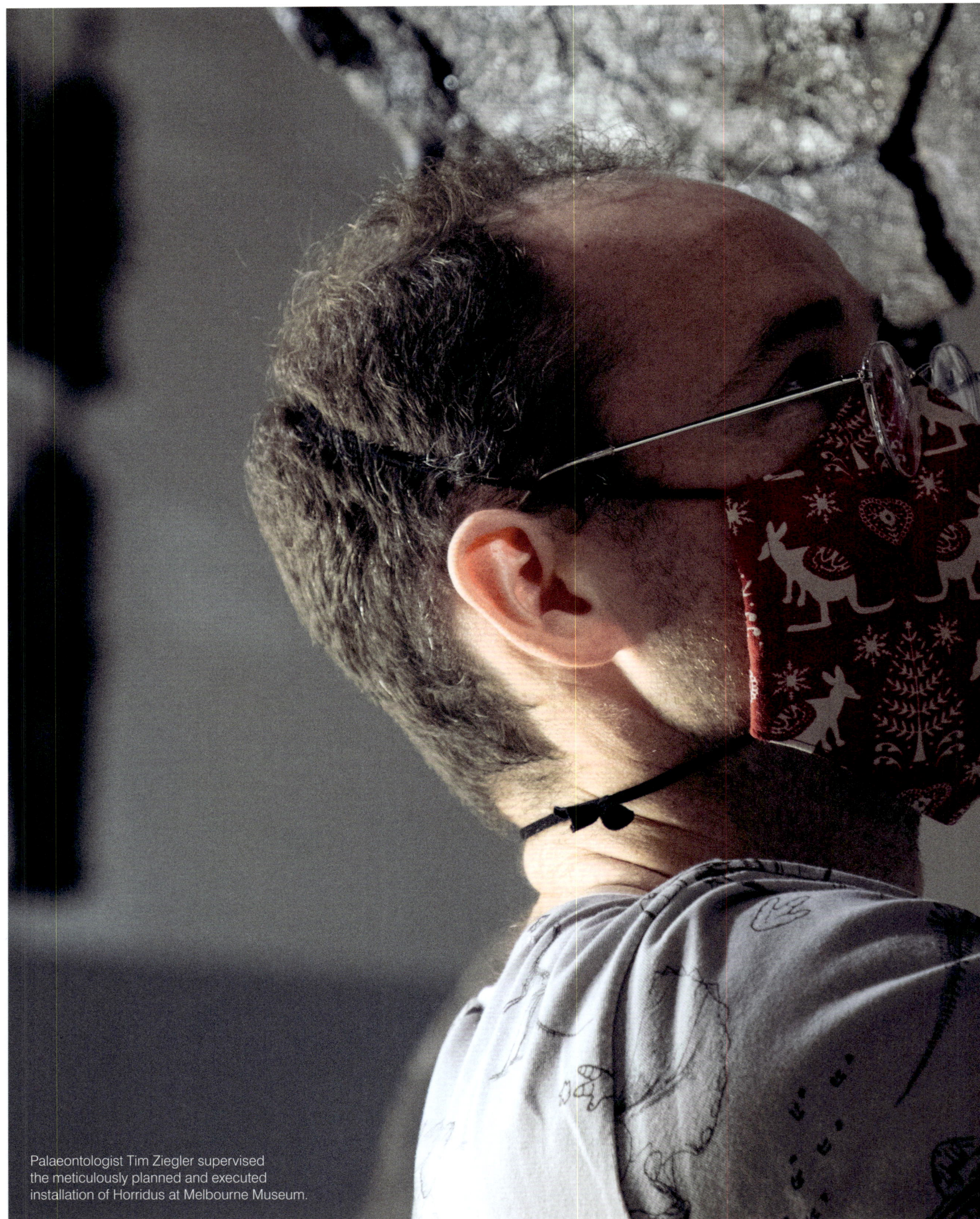

Palaeontologist Tim Ziegler supervised the meticulously planned and executed installation of Horridus at Melbourne Museum.

The stone matrix containing the fossil was encased within a field jacket that enabled its transport from Montana to Vancouver Island.

Philip J. Currie

Professor + Canada Research Chair,
Dinosaur Palaeobiology,
University of Alberta

I discovered dinosaurs as a six-year-old when I found a plastic model in a box of cereal. It was part of a series of eight prehistoric animals that were meant to entice kids like me to get their parents to buy that cereal until they had the whole set. I was hooked and managed to get every dinosaur in the series (including *Triceratops*) except for the one I really wanted—the *Tyrannosaurus rex*. I was dinosaur crazy from then on, although I don't think I thought about becoming a palaeontologist until I was 11 years old and read *All About Dinosaurs* by Roy Chapman Andrews. His descriptions of fieldwork, research, travel and dinosaurs (of course) set me on my career path the day I read it. Within a year or two, I learned from visiting the dinosaur displays at the Royal Ontario Museum in Toronto that many specimens were excavated in Alberta. So now I knew I wanted to look for dinosaurs and live in Alberta, and I set out to do exactly that. The miraculous thing is that I hung onto that goal throughout my schooling, and at the age of 27 I became the curator in charge of collecting dinosaurs at the Provincial Museum of Alberta!

Over my career, I have focused most of my research efforts on two groups of dinosaurs—the theropods and the ceratopsians. The apices (most derived, largest and last) of these two types of dinosaurs were respectively *Tyrannosaurus* and *Triceratops*. Although both dinosaurs were originally found in the United States, they also lived in Canada, with specimens found in Alberta and Saskatchewan. To me it is fascinating that such large and derived dinosaurs, representing millions of years of evolutionary change in their respective lineages, constitute the end of the age of dinosaurs.

There is no question that *Triceratops* can inspire awe by its size alone. However, the realisation that this was one of the most sophisticated plant-eaters in the world more than 67 million years ago inevitably leads us to speculate on the details of its life. Did *Triceratops* take care of and protect its young? How did it change its appearance as it grew up? And how long did it live? Were those enormous horns used to fight off *Tyrannosaurus rex*? Research provides the answers to some of these questions, but every answer leads to the development of more questions.

Virtually every dinosaur skeleton that we collect provides new information about that species of animal, and the better preserved the specimen is the more likely we are to learn something unique. Over the years we have moved from studying the gross features of dinosaur skeletons to looking at smaller details preserved in the fossils. This amazing skeleton, for example, is so well preserved that we can look at microscopic cross-sections of the bones to determine how old it was when it died or study the chemical composition of the bones to learn what it was eating when it was alive.

Dinosaurs like this one are never found in isolation. They are found in association with plant fossils and the bones of animals that lived in the same environment—from tiny mouse-like mammals to giant meat-eating dinosaurs like *Tyrannosaurus rex*. Each of these fossils, along with the rocks they are found in, provide information on the environment of that time. And years of research in the same fossil beds gives us a sense of how much biodiversity there was in the local area when the *Triceratops* was alive.

The thing that excites me most about this *Triceratops* skeleton is its completeness and the quality of its preservation. We learn a lot about how dinosaurs changed their proportions as they grew up from complete and nearly complete skeletons. And although many *Triceratops* specimens have been collected since the first one was described in 1889, believe it or not there is a paucity of specimens that include both the skulls and skeletons. Furthermore, the preservation of the bones on this specimen is so exquisite that we can look at the courses of nerves and blood vessels across the surface of the skull, or where muscles attached to the bones.

Some ribs on the *Triceratops* were extremely long and had to be assembled from pieces.

> The preservation of the bones on this specimen is so exquisite that we can look at the courses of nerves and blood vessels across the surface of the skull, or where muscles attached to the bones.
>
> Philip J. Currie

Philip J. Currie, one of the world's foremost palaeontologists and dinosaur experts, inspected and assessed the *Triceratops* specimen in Canada.

As a large animal weighing about the same as a bull elephant, *Triceratops* would have had a difficult time surviving today because of the wholesale destruction of the kinds of environments that it would have preferred. Big animals need big home ranges, and developments like farming, roads and cities have dissected those ranges so that big animals are no longer able to live in the large areas that they require to migrate, find enough food throughout the year and so much more. Above all else, hunting pressure would render the survival of *Triceratops* difficult or even impossible.

I study many aspects of dinosaur anatomy for many different reasons. The braincase is amazing because in a dinosaur you can get a sense of what parts of the brain are better developed than others, and you can see where nerves radiated out of the braincase to innervate and control the body when that animal was alive.

I hope the *Triceratops* exhibition at Melbourne Museum inspires people to ask questions and learn from the lessons of natural history. Dinosaurs dominated the world for 150 million years during the Mesozoic, but it did not save them from a catastrophic demise 65 million years ago. Nevertheless, dinosaurs did bounce back because birds are directly descended from carnivorous dinosaurs similar to *Velociraptor*. Under modern biological or palaeontological classification systems, birds are part of the Dinosauria. And because of that, we have more than 10,000 species of living dinosaurs, which is more than double the diversity of living mammals. No matter how you look at dinosaurs, there are so many reasons we should be humbled by their history and presence.

Close-up shot of the ossified tendons.

The discarded stone matrix surrounding the fossil was collected as the specimen was removed from its field jacket.

> To me it is fascinating that such large and derived dinosaurs, representing millions of years of evolutionary change in their respective lineages, represent the end of the age of dinosaurs.
>
> Philip J. Currie

NMV P 256878.160

The left and right femurs, from the hind legs.

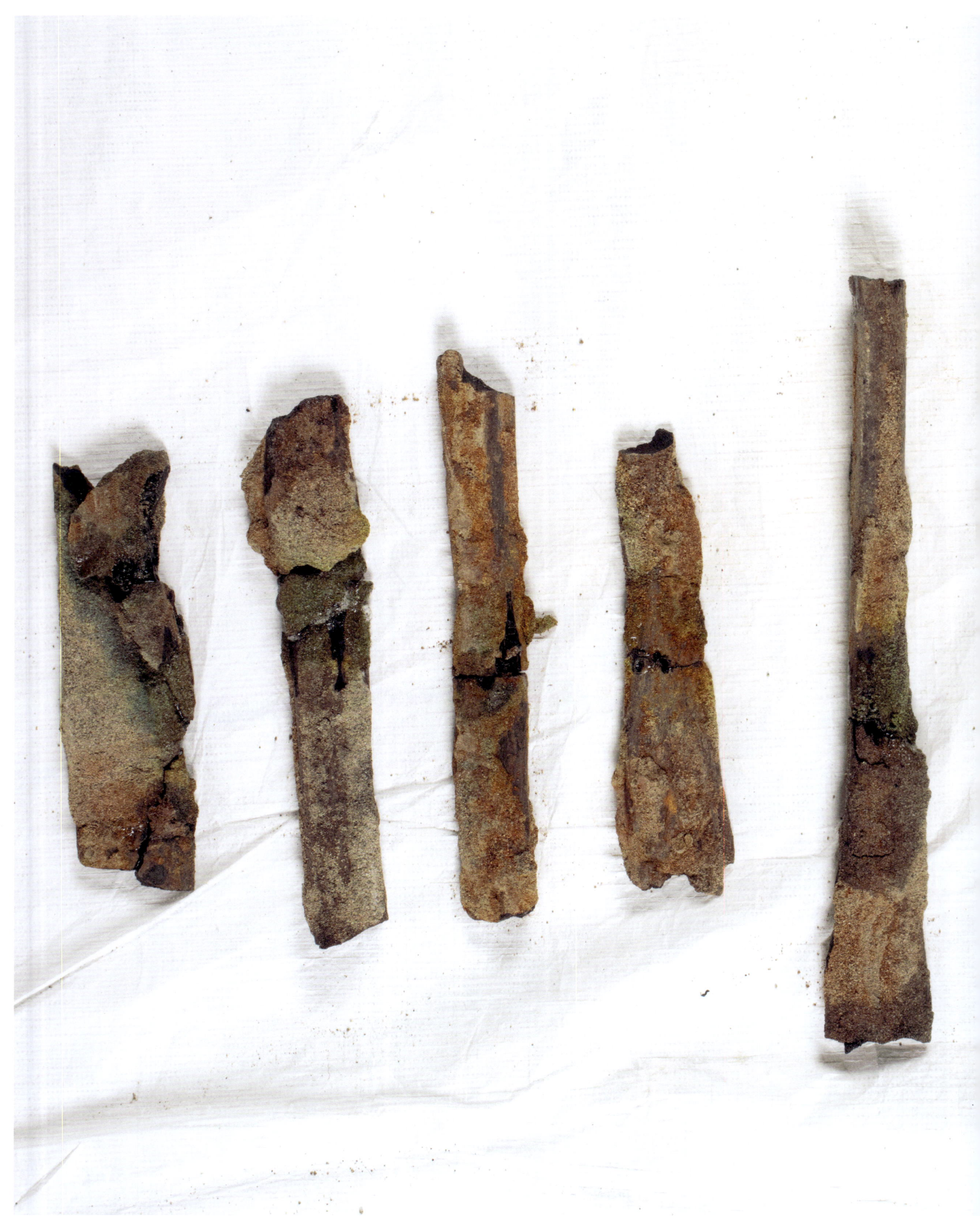

Elaine Vallis

Fossil Restoration Technician
Dino Lab

It was a jigsaw puzzle trying to figure out how to get the bones in the crates. Once it was done, I looked at them and thought, *Oh, there's a dinosaur in there!* It was like the weirdest IKEA furniture of all time.

The best moment was seeing the skeleton mounted. Prior to that, I was unable to comprehend how big it was. There were lots of super-tiny bones that went into this massive animal. Realising that this was a real creature was amazing. And there I was sitting underneath it.

A series of isolated ossified tendons, detached from the vertebrae.

Horridus was packed into crates for air transport between Canada and Australia.

DINO
LAB
PTR0159
1238KG

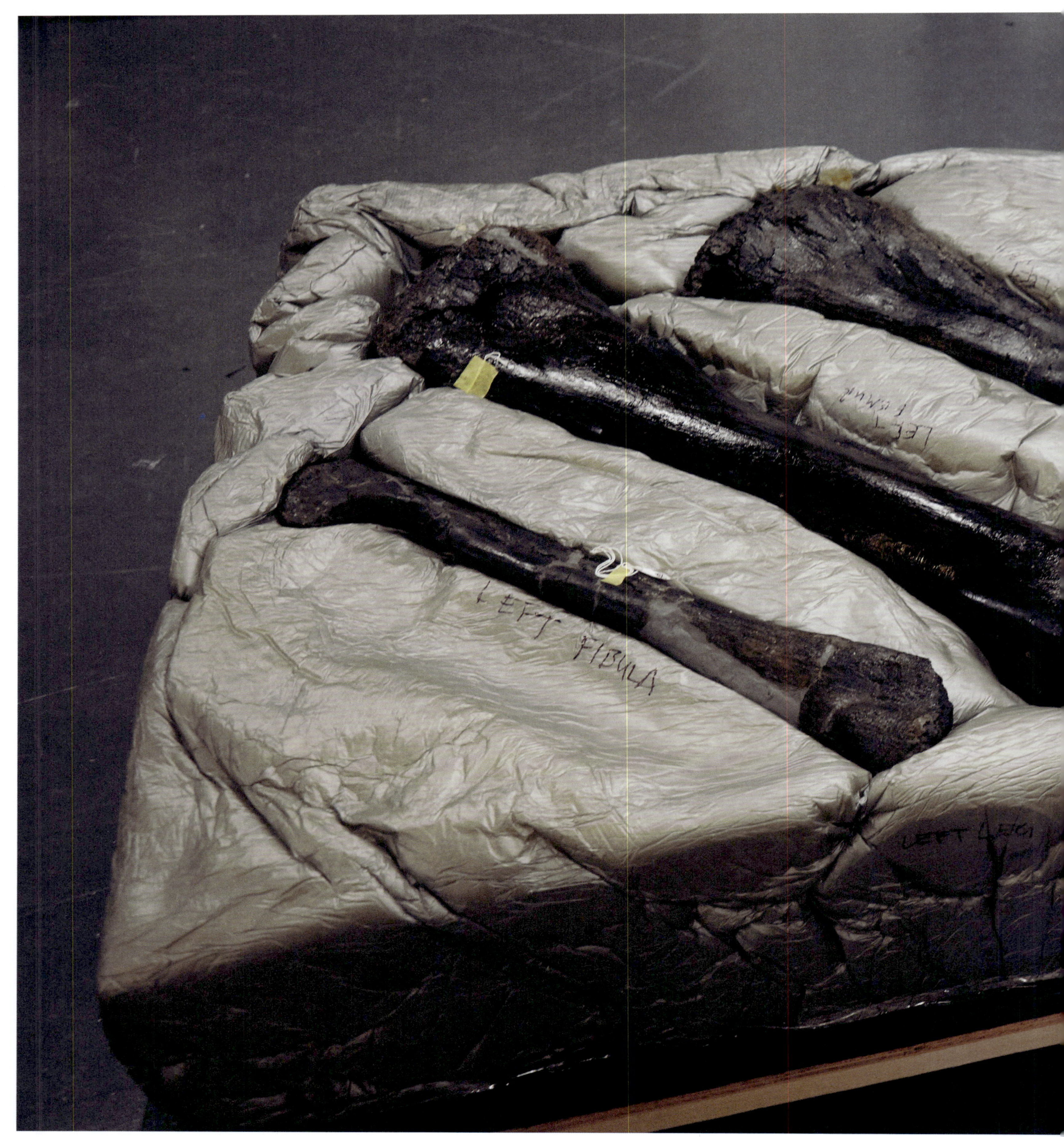

When unpacking those crates, staff at Melbourne Museum were able to witness the extraordinary care that had gone into transporting a dinosaur between continents.

> It was like the weirdest IKEA furniture of all time.
>
> Elaine Vallis

Close-up shot of the left humerus (upper arm bone).

Nancy Ladas

Manager, First Peoples Collections
Museums Victoria

When Horridus was packed in crates by the Dino Lab team, waiting for their paperwork to be finalised, for fun I started sending messages to my colleagues from the dinosaur. Saying, 'I'm packed up here, I can't say much, I'm just waiting to find out about the next part of my journey.'

I cried in the loading dock when the truck arrived in Melbourne. Every crate that we opened revealed a layer of bones and the colour was magnificent—a beautiful, shiny, dark chocolate colour. It was magical. I'm glad that I participated in the unpacking because I would have been one of the first people to see the fossil in all its scope.

I felt like I was travelling with the fossil because I spent so much time organising its flight. When we finally unscrewed the top of the crate in Melbourne, I thought they must be so glad to get some fresh air. There were layers and layers of spray foam surrounding the bones. It was like a car model kit. I realised then that someone was going to have to put it together! It was exciting to see each crate be cracked open and life spring from the foam.

We recreated some of the bones that are missing out of a putty. When Dino Lab shipped everything from Canada, they left the fake bits on the metal armature, so it looked like a sad Christmas tree when we first unpacked it.

The skeleton is 1000 kilograms of real bone. You need a couple of people to manoeuvre it. The knuckle bones are cute—they're kind of like ours, but chunkier.

Every week when we met with Terry Ciotka from Dino Lab, we needed to know which bones they'd taken out of the rock and the matrix. We needed to know what they had discovered was missing. We had three tables and we'd say, 'Okay, how much of this bone have you uncovered, and what percentage complete is it?' I had no idea how to connect the names of bones with the shapes I was seeing.

This dinosaur is alive for me. I manage the Australian and international First Peoples collections. During lockdown, I went in once or twice a week to visit the ancestors. It bothered me that they were sitting there in the dark. I would say hello, explain what was happening in our world, switch the lights on and open the door so the ancestors could breathe. The whole reason for a museum to exist is to bring what we have to life for the public—to connect with the living past. I want to do the same with this creature. We have to honour this animal, their life, their journey.

The phalanx is part of *Triceratops*'s complex feet.

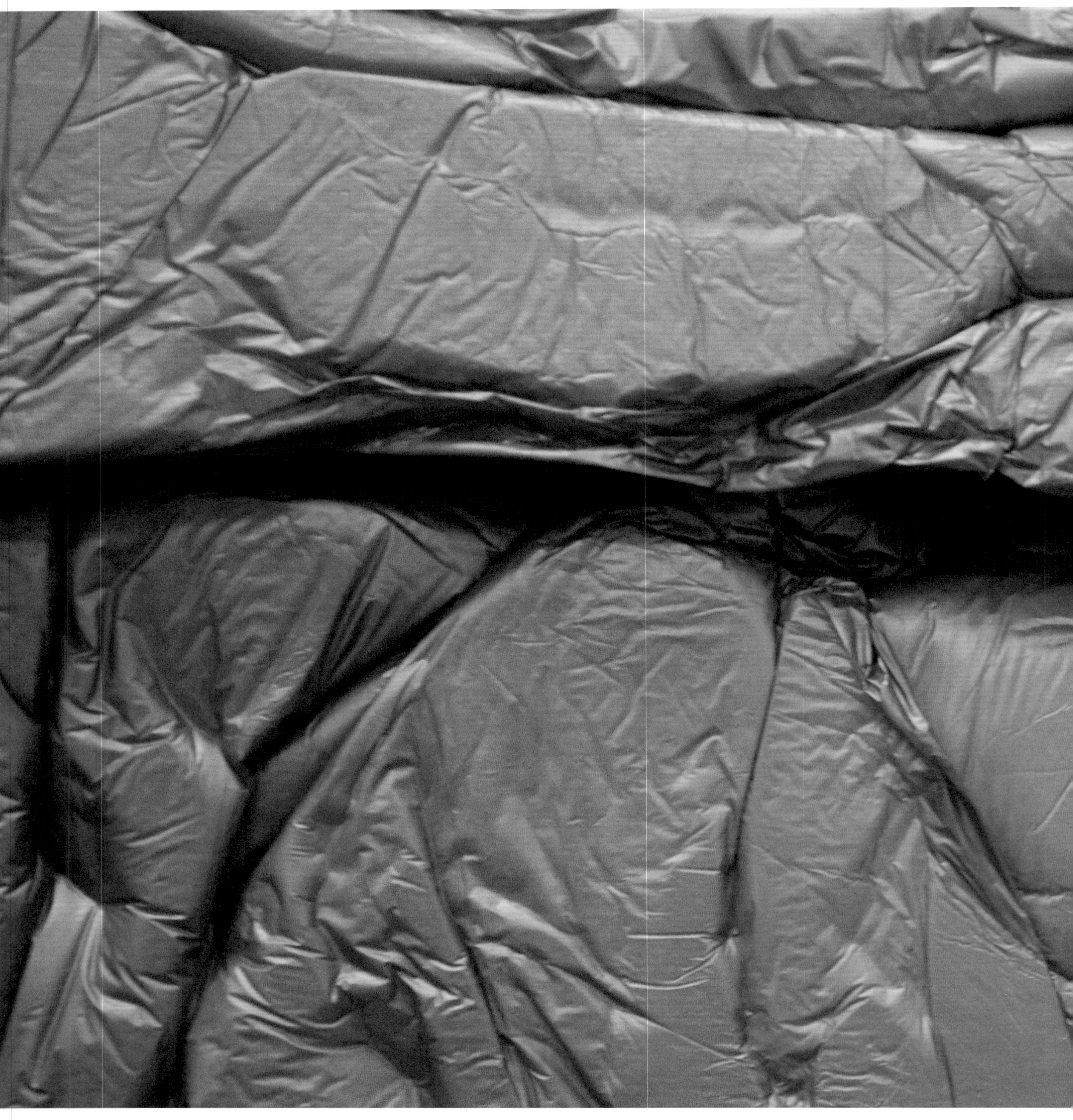

The fragile fossil was packed inside layers of cushioning foam.

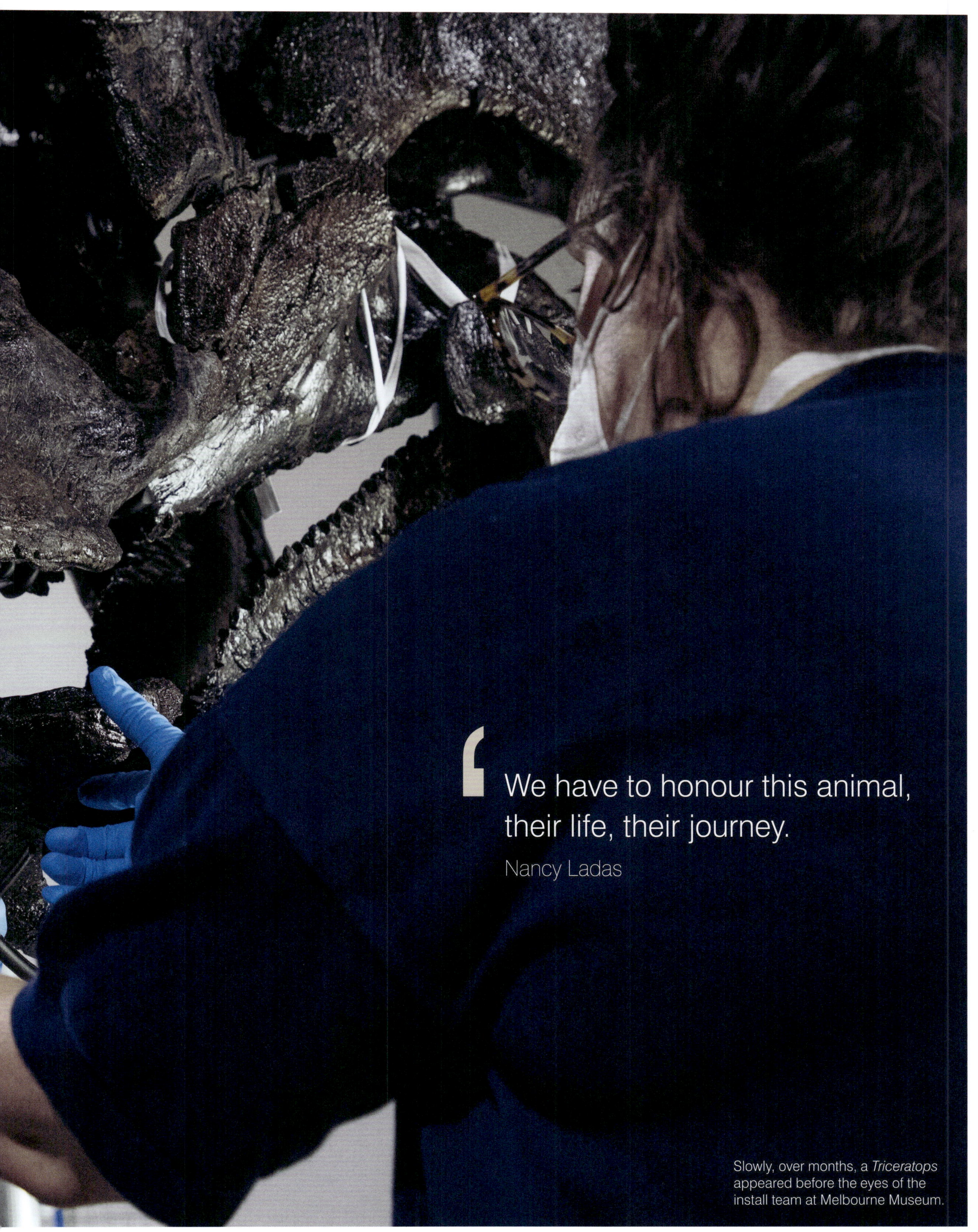

> We have to honour this animal, their life, their journey.
>
> Nancy Ladas

Slowly, over months, a *Triceratops* appeared before the eyes of the install team at Melbourne Museum.

Bones were held securely in place during transit by form-hugging packaging, sleeping through the long flight.

Sometimes six people were required to manipulate a single piece of fossilised bone.

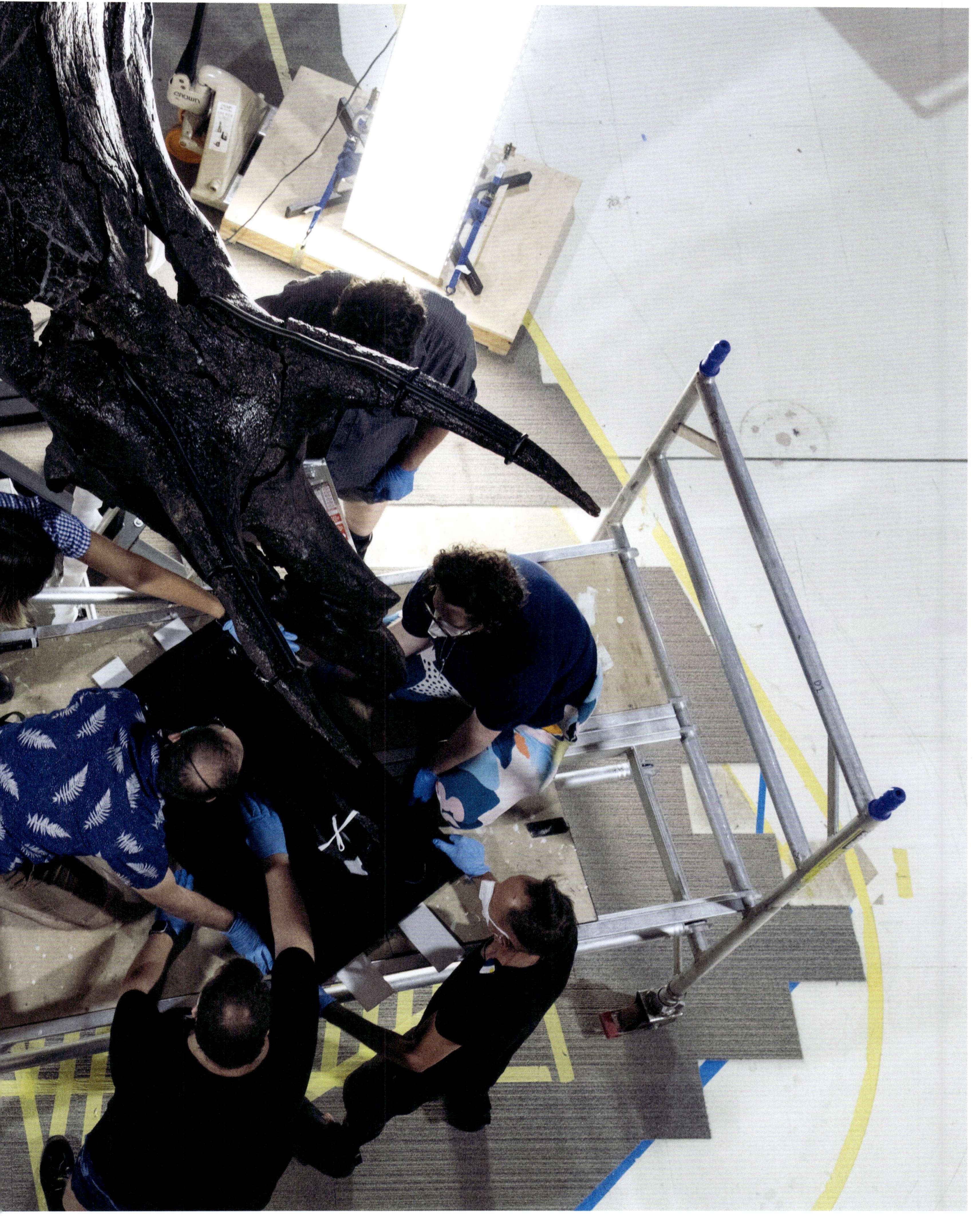

Dr Linden Ashcroft

Science Communication
University of Melbourne

Triceratops is one of those dinosaurs that is instantly recognisable. Neither my toddler nor I know our *Ankylosaurus* from our *Ichthyosaurus* yet, although both creatures feature in picture books currently on high rotation at bedtime. But I know that when we see *Triceratops* at the museum, my daughter (and probably I) will point and squeal with recognition.

Seeing or touching even one part of a creature that was alive millions of years ago can be a life-changing experience. It reminds us that we have only been on Earth for a small fraction of its existence, and that it had seen so many things before we came along and made such a mess. I'm also struck by how the planet would have been during *Triceratops*'s lifetime. The atmosphere, the ocean, the land and even the stars would have looked different then. It's both humbling and comforting to think that although things are quite different now, we've shared the same Earth as this creature.

I used to think of extinction as something that happened suddenly to the poor old dinosaurs, with visions of meteors and spewing volcanoes, and the fossil evidence of creatures perishing while running away through the mud. It seemed instant, violent and loud. But the extinctions we are experiencing on Earth now are much more silent. They are no less violent, and still very sudden on a geological scale. It makes me wonder what evidence is being left behind of the plant, insect and animal species we are currently losing as our planet warms.

The occipital condyle is a rounded piece on the rear of *Triceratops*'s skull where the first neck vertebra connects the head and body.

> If a tram is thirty rhinos, then a *Triceratops* must be at least half a packed tram full of fans coming home from the MCG. With horns. Imagine that charging around the Hoddle Grid!
>
> Dr Linden Ashcroft

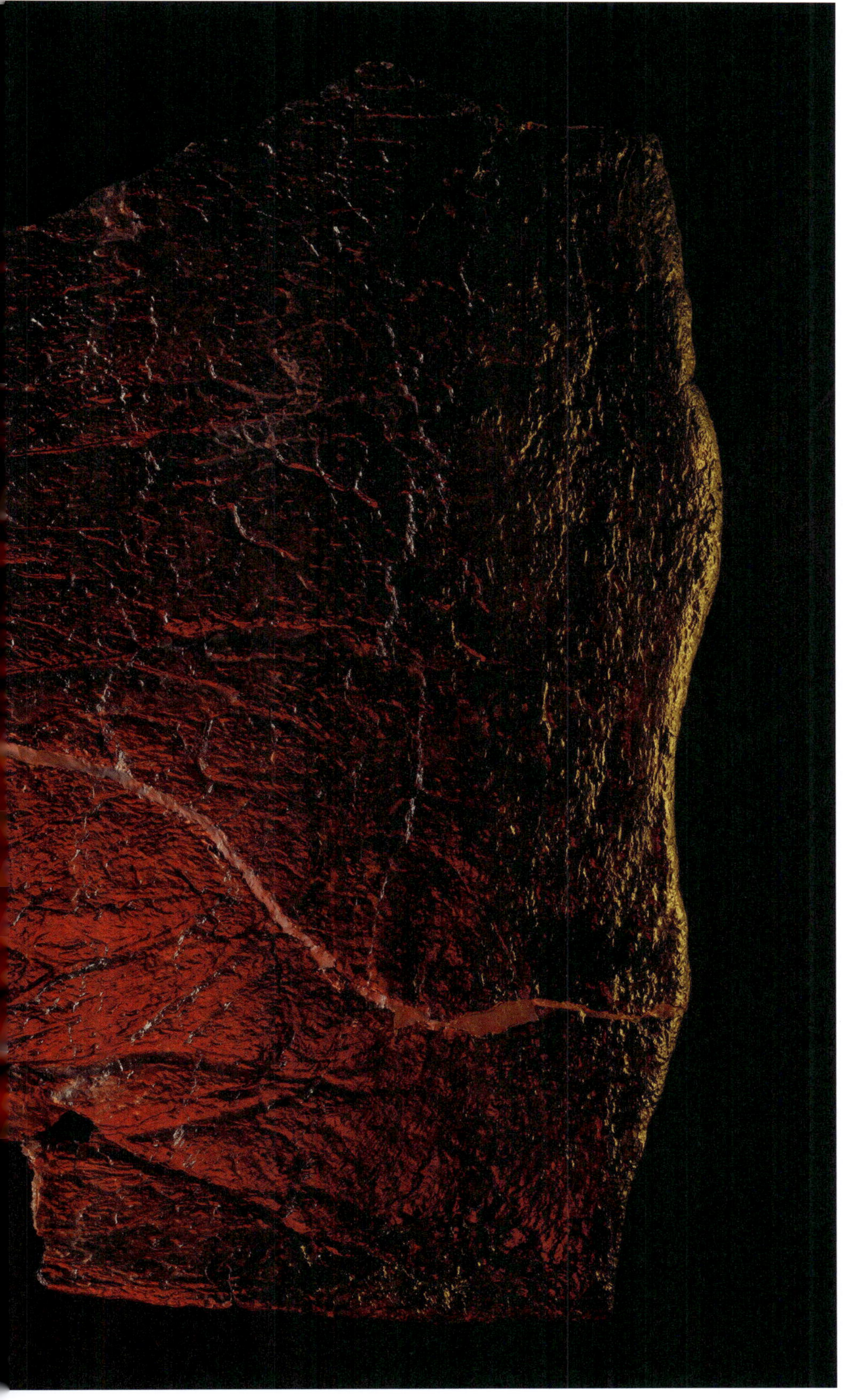

Having such a complete *Triceratops* skeleton here, in Melbourne, is thrilling for our city. Just think: every time you walk, play or picnic in Carlton Gardens, you are metres away from one of Earth's most spectacular creatures. It's also exciting and humbling for Melbourne to be a part of this skeleton's story: every scientific artifact in Melbourne Museum has made an incredible journey to get here, but the 67-million-year adventure for this *Triceratops* is something truly special.

A *Triceratops* would make an absolute mess of today! If a tram is thirty rhinos, then a *Triceratops* must be at least half a packed tram full of fans coming home from the MCG. With horns. Imagine that charging around the Hoddle Grid!

Triceratops are often painted as the friendly fellows of the dinosaur world. Yes, they've got terrifying horns and that formidable spiky frill, but they don't eat other dinos—they eat leaves and grass! So why the armour? Protection? Romance? To secure the most tender green sprouts? I get lost imagining the world *Triceratops* inhabited—a world where such impressive armour was needed for a grazer.

Dinosaurs and fossils are a gateway to science for many people. They were big, cool and scary: what's not to love? Many types of science are involved in answering big dinosaur questions. When they were around, how they lived, how they moved, how they died and what world they lived in—these questions require physics, chemistry, earth sciences, physiology, ecology and many other fields. My hope is that *Triceratops* becomes a majestic gateway for Victorians to become more connected with science that helps us not only understand the past, but the future too.

A fragment of *Triceratops*'s iconic frill.

Hazel Richards

Curatorial Research Assistant, Palaeontology
Museums Victoria

No other land animal has a head this big and heavy. The frill is solid, thick bone. It's absolutely bonkers thinking about how this animal might have moved. We deliberated and struggled with trying to decide how to put this creature together in a way that made it look like a realistic, dynamic, moving animal captured mid-stride while still conveying that enormous bulk and giving it a sense of mass and weight.

In most dinosaurs the centre of gravity lies just in front of the hips, but with a big-headed animal like *Triceratops* it's going to sit way forwards relative to the rest of the body. And yet when you look at its front legs, which are presumably taking more of its weight proportionally, they're not as robust as you might expect. In developing the posture, we're thinking about how it would have borne its weight across its four limbs as it took each step. That's also informed by the peculiarities of the fossil itself. One of the key things that we found out as the fossil was being prepared is that there's an interesting left/right asymmetry. The femurs are remarkably different in length. We think this is a process that happened during fossilisation—bones became distorted due to pressure and heat and rocks pressing down on them for millions of years. What we're faced with is an animal that was probably symmetrical in life, but the fossil is lopsided. When mounting the skeleton, we wound up exaggerating the spacing between each bone so that its shorter right leg was spaced out more to give it a relatively equal length between both legs. We decided on a pose that incorporated that length difference while disguising it as much as possible.

This specific peculiarity about the skeleton informed decisions we made about how it went up on the mount, because once you make the knee joint on one side really big, you have to make that same decision everywhere else or it's going to be inconsistent. This required lots of tweaks, as solving one problem has a knock-on effect in the way that the skeleton fits together everywhere else. And that's just dealing with it virtually. When it comes to the artisans creating the steel armature that Horridus physically sits on in real life, they have to fit all of the brackets and bolts together in there too. The physics and constraints of putting these big bones up on an armature adds another dimension of complexity. We had to find a compromise between putting the skeleton back together how it was found in the ground versus reconstructing an animal as it would have been in life, but we also needed to make something that is mechanically stable from an engineering standpoint and fits within the constraints of the physical building that it's going to stand in for decades.

Very little of the skeleton was broken. While the joints might have separated from where they were in its death position, when you reconstruct the skeleton in 3D the way that it was found in the ground it looks like this animal simply lay down and went to sleep. The rib cage is still rounded and intact—you can see the whole body cavity inside. Putting it up on the mount is like we're pulling the marionette strings and bringing Horridus back to life.

To create the 3D scans, we used a handheld device with an LED and lens array that projects patterned light onto the object and converts distortion of this light into a 3D shape. It builds up the shape of the specimen as we swipe the scanner in space around it, while also taking many simultaneous photographs of the fossil's surface texture. All the colour and texture variations on the surface are recorded by the camera—even the numbers that we've written on the fossil to catalogue it in our collection are captured. Another method we've been using to image the *Triceratops* is photogrammetry. This is similar to surface scanning but uses a traditional camera. We photograph at regular intervals through 360 degrees, from various angles, and then the computer can use these hundreds of images to reconstruct the surface in 3D. The final way is CAT scanning, which is essentially a three-dimensional X-ray. This allows us to reconstruct the inside of the fossil as well as the outside. This is really cool for something like the braincase, the part of the skull containing the hollow cavity that the brain occupied in life. Put that through the CT scanner and you have the negative space where the brain used to be, allowing you to build a three-dimensional shape that represents the brain of an animal that's been extinct for 67 million years.

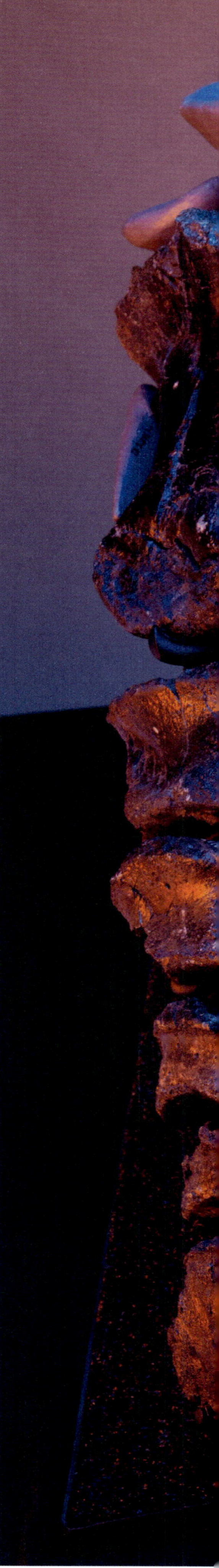

The rear right foot (pes).
Replacements for missing bones are in grey.

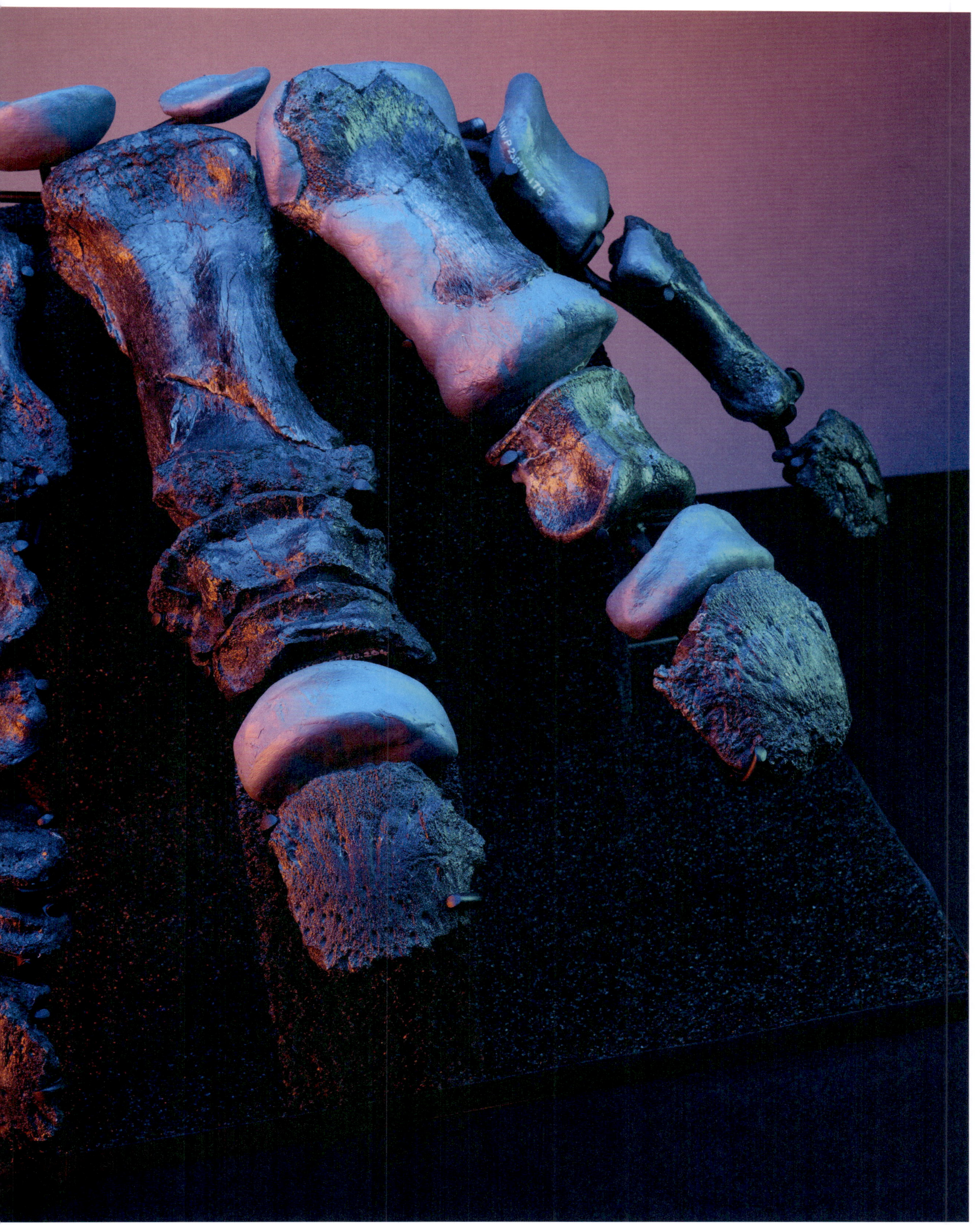

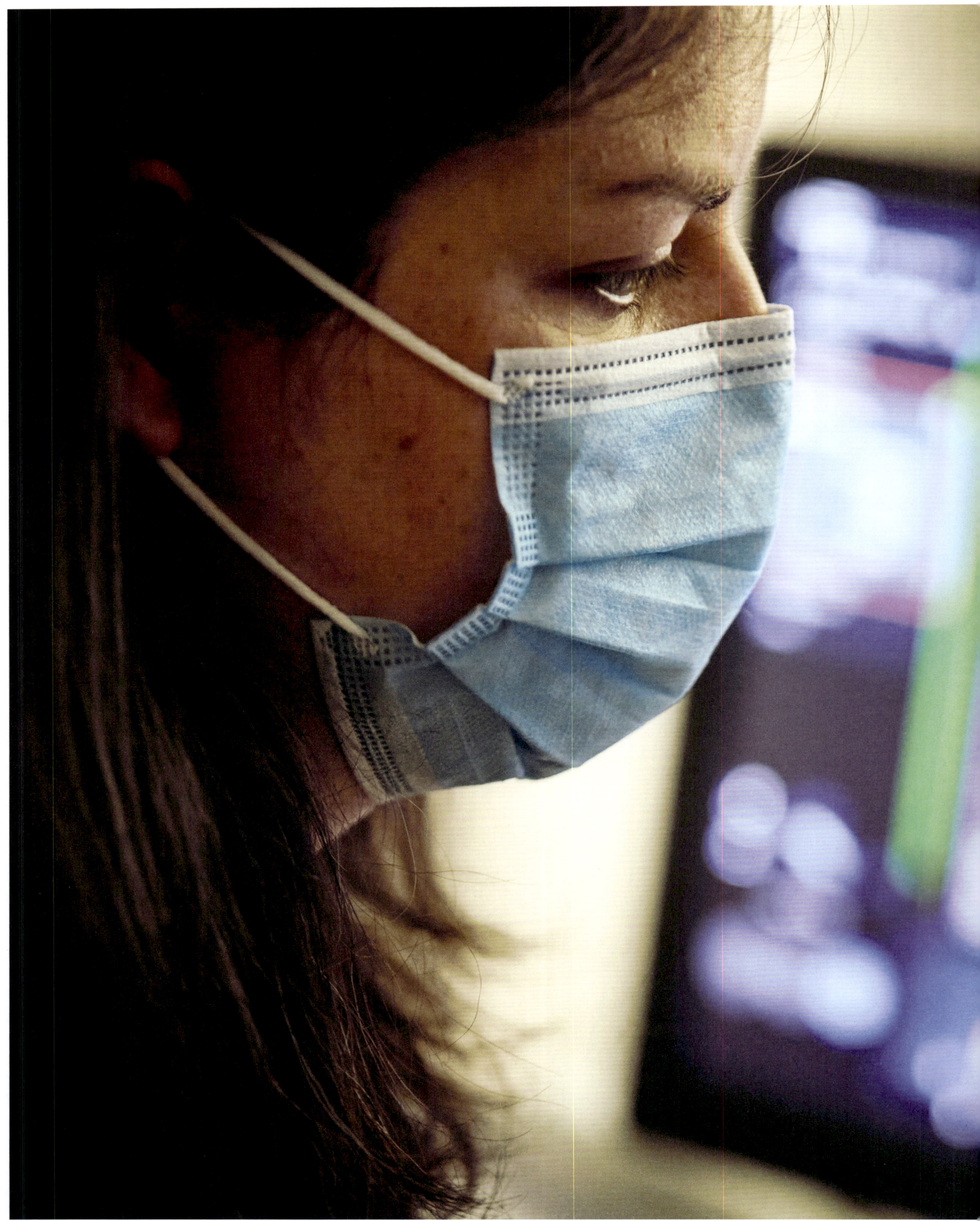

Hazel Richards runs the CT scanner, gathering primary data on the *Triceratops*.

Having worked with the work-in-progress 3D scans from Canada for a year by the time the specimen arrived, it was breathtaking to see the fossil for the first time. When you look at it in the computer, you lose all sense of scale—of how big these animals are. In real life, it has a presence, almost like it's buzzing, like it has an aura. I was not prepared for that when I walked in there and encountered Horridus for the first time.

I'm a functional anatomist, so I think about the way animals are structured and how they move in life. There are fundamental constraints imposed upon all land-dwelling animals. When you are a big animal moving around on the earth, gravity acts on you in the same way today as it did on dinosaurs millions of years ago. Even though *Triceratops* is very different to a rhinoceros, we can use information from living animals to understand how animals like *Triceratops* were put together, how they moved, how they faced those challenges of holding up their body weight, moving around, how they found enough food to sustain such a large organism and how they communicated with their mates or rivals. There are echoes across time of these same solutions that evolution comes up with to address these problems. 'Nature' is more poetic, but if you're comfortable with it 'evolution' is slightly more precise.

You can depict *Triceratops* facing off with a *T. rex*, locked in battle for all time, but in reality 99% of its day would have been spent eating.

The hands, or manus (front feet), are made up of dozens of little bones in the wrist, palm, fingers and fingernails (or hooves). In *Triceratops*, the front and back feet are very different to one another. When I put the hands together for the first time using the 3D models, I saw how placing those joint surfaces in articulation created an amazing arc across the hand. That curvature beautifully matched the way the forearm bones fit together. It made sense in terms of thinking about how the weight went down through its foreleg, through its hand and into the ground. It was totally different to any other animal I'd ever worked on. Having to start from that hand and work back meant revising the way that the whole animal was put together based on the fact it was bearing all this weight on three little toes on its front foot. It was a moment of revelation.

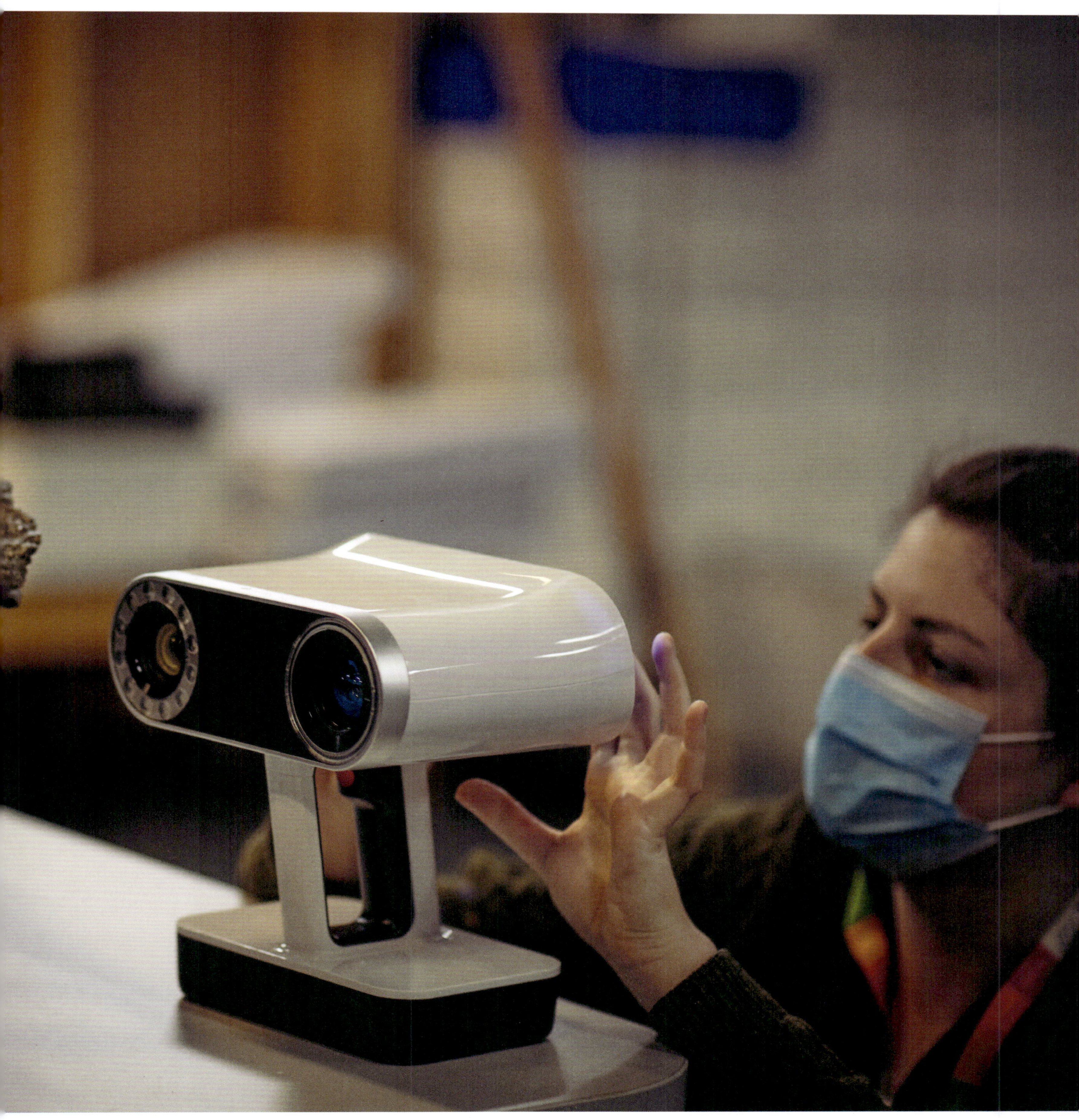

Hazel Richards using a handheld surface scanner to capture 3D scans of Horridus.

Elements of the skeleton were remarkably intact, fitting together like a jigsaw puzzle.

A hoof bone (ungual phalange), from the tip of one of Horridus's left fingers.

> In real life, it has a presence, almost like it's buzzing, like it has an aura. I was not prepared for that when I walked in there and encountered it for the first time.
>
> Hazel Richards

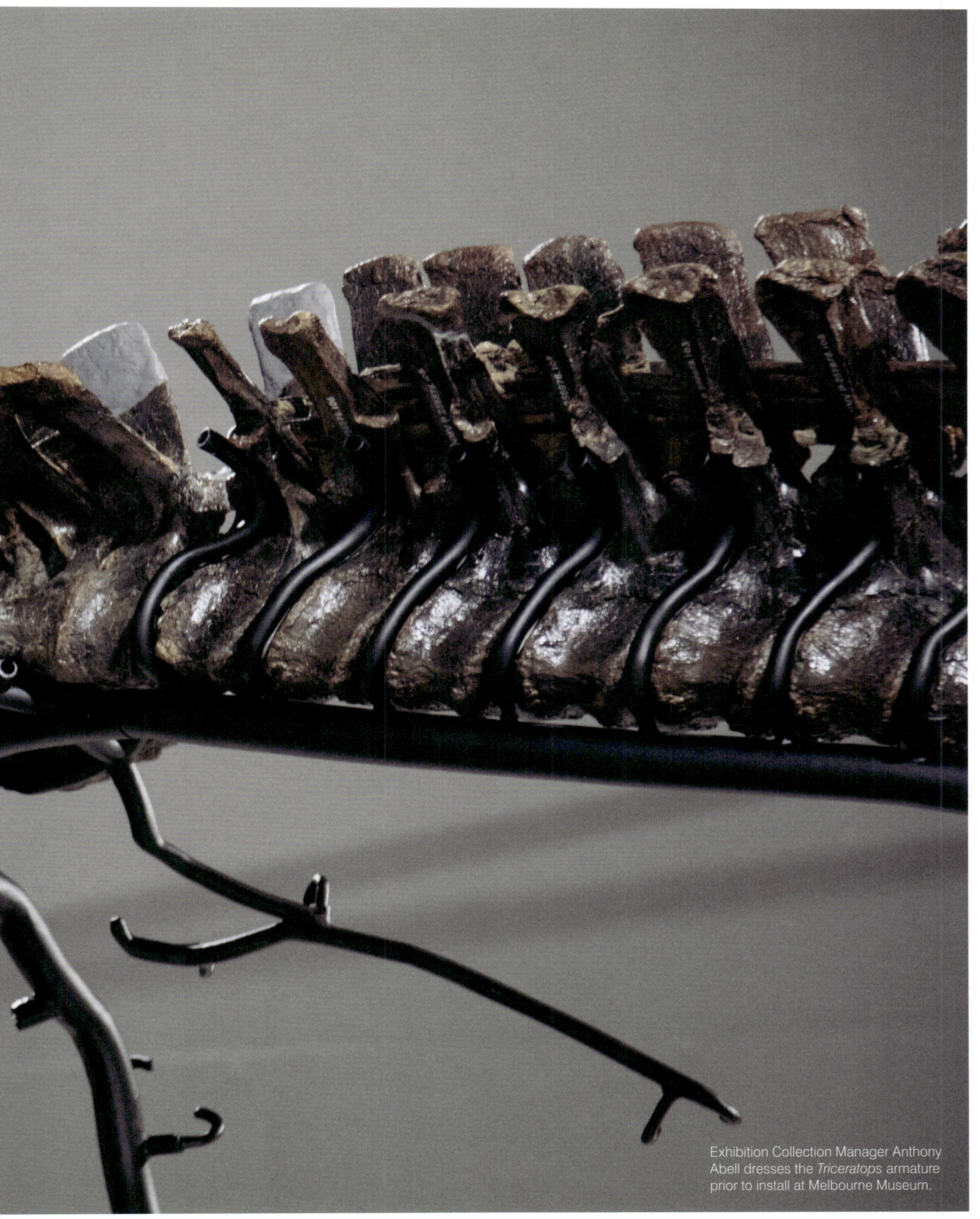
Exhibition Collection Manager Anthony Abell dresses the *Triceratops* armature prior to install at Melbourne Museum.

After many months of methodical construction, Horridus is nearly ready for display in the exhibition space at Melbourne Museum.

CONTRIBUTORS

Benjamin Law is a writer and broadcaster. He is the author of *The Family Law*, which was adapted into an award-winning SBS TV series, and is a co-creator, producer and co-writer of the upcoming Netflix show *Wellmania*.

Blair Holtner is a cabinet maker and lead fossil restoration technician at Dino Lab.

Chris Flynn is the author of *Mammoth, Here Be Leviathans* and *Horridus and the Hidden Valley*. He is Editor-in-residence at Museums Victoria.

Craig Pfister is a commercial palaeontologist. He discovered the *Triceratops* on private property in Carter County, Montana. He noticed part of the pelvis protruding from the sandstone and from there Horridus was unearthed. Craig began excavation in September of 2014 and the site was completed by October of 2015.

Dermot Henry is Head, Sciences Department at Museums Victoria. He has published articles on a variety of topics covering minerals, rocks and meteorites, and has contributed to four books on Victorian minerals.

Elaine Vallis is a fossil restoration technician at Dino Lab.

Erich M. G. Fitzgerald is Senior Curator of Vertebrate Palaeontology at Museums Victoria. His specialty is in the evolutionary history of marine vertebrates, especially marine mammals.

Hazel Richards is a curatorial research assistant in palaeontology at Museums Victoria. She specialises in anatomy and her main field of interest is extinct marsupial megafauna.

Professor John A. Long is Strategic Professor in Palaeontology at Flinders University in Adelaide. He was previously Vice President of Research and Collections at the Natural History Museum of Los Angeles County and is the author of more than 30 popular science books.

Laura Jean McKay is an author and creative writing lecturer. Her novel *The Animals in That Country* won the Victorian Prize for Literature and the Arthur C. Clarke Award.

Dr Linden Ashcroft is a lecturer in climate science and science communication at the University of Melbourne. She is a Science and Technology Australia Superstar of STEM, a program that challenges gender assumptions about scientists.

Maree Clarke is a multi-disciplinary artist and pivotal figure in the reclamation and promotion of southeast Australian Aboriginal art practices. Her work is the focus of a major career survey, Ancestral Memories, at the National Gallery of Victoria.

Mark Radburn is a cast member on the ABC and Netflix TV show *Love on the Spectrum*. He is a presenter for Autism Awareness Australia.

Melissa Kay is an artist and fossil restoration technician at Dino Lab.

Mitch Mahoney is a Boon Wurrung artist and cultural educator with a passion for marine biology and transdisciplinary creative practice. Recent exhibitions include *Weaving Sustainable Culture* at Footscray Community Arts, and Science Gallery Melbourne's *DISPOSABLE Eel Trap*, installed on the Maribyrnong River.

Nancy Ladas is Manager, First Peoples Collections at Museums Victoria.

Philip J. Currie is a Canadian palaeontologist and museum curator. He helped found the Royal Tyrrell Museum of Palaeontology in Drumheller, Alberta and is a professor at the University of Alberta in Edmonton. He is an editor of *Encyclopedia of Dinosaurs* and one of the world's foremost experts in tyrannosaurs.

Rebecca Dart is a Vancouver-based art director and designer working in the animation industry. She is one of the world's most popular paleoartists.

Ry Williams is the lead welder, fossil restoration technician and dinosaur levitation expert at Dino Lab. He created the armature on which Horridus is mounted in Melbourne Museum.

Shannon Faulkhead is Head, First Peoples Department at Museums Victoria. She is the former Chief Executive Officer of the Koorie Heritage Trust and an author, academic, cultural critic and practitioner.

Shannon Martinez is the owner and chef at plant-based businesses Smith & Daughters, and Smith & Deli. She has been a guest judge on *Masterchef Australia* and is the author of four books, including *Smith & Daughters: a cookbook (that happens to be vegan)*.

Shaun Tan is an artist, writer and filmmaker. He won an Academy Award for his short film *The Lost Thing*. He is a recipient of the Astrid Lindgren Memorial Award and the Kate Greenaway Medal.

Dr Susannah C. R. Maidment is a palaeontologist at the Natural History Museum, London. She is renowned for her research on ornithischian dinosaur evolution and is an authority on stegosaurs.

Terry Ciotka is co-founder and Operations Manager at Dino Lab, a Canadian fossil restoration lab, educational destination and prehistoric boutique in Victoria, British Columbia. His team acquire and prepare fossils for exhibition at museums across the globe.

Tim Flannery is an author, palaeontologist, explorer and conservationist. He has discovered more than 30 mammal species, is leader of the Climate Council and was Australian of the Year in 2007.

Tim Ziegler is Collection Manager, Vertebrate Palaeontology at Museums Victoria. His specialty is taphonomy, the traces left behind on fossil bones, such as bite marks, weathering and geochemical changes.

Thomas H. Rich is Senior Curator of Vertebrate Palaeontology at Museums Victoria. He has published 12 books and more than 100 scientific papers. He leads the field in research related to the polar Cretaceous tetrapods of Victoria.

Tony Birch is the author of 10 books, including *The White Girl* and *Dark as Last Night*. In 2017 he was awarded the Patrick White Literary Award for services to Australian literature.

IMAGE CREDITS

Pages: 6, 10, 16, 39, 42, 101, 110, 130, 140, 144, 166, 178, 184, 189, 198, 210, 214
John Broomfield

Pages: front cover, 8, 18, 32, 61, 62, 64, 68, 72, 80, 96, 102, 106, 122, 124, 127, 136, 150, 196, 204, 220, 222, 224, 230, 234, 239, 240
Benjamin Healley

Pages: 22, 23, 28, 30, 31
Craig Pfister

Pages: 20, 24, 26, 78
Heinrich Mallison

Pages: 35, 36, 44, 48, 52, 58, 66, 70, 74, 82, 84, 86, 88, 90, 98, 104, 108, 116, 120, 134, 142, 146, 148, 152, 154, 156, 158, 162, 164, 168, 171, 172, 174, 180, 186, 190, 193, 194, 218, 228, back cover
Trevor Dixon Bennett, Kingtide Films

Pages: 13, 40, 50, 56, 76, 114, 132, 139, 176, 177, 202, 207, 208, 212, 216, 227
Jon Augier

Pages: 46, 54, 113, 118, 139, 160, 178, 192
Dino Lab

Pages: 92, 94
Rebecca Dart

Page: 128
Shaun Tan

Page 14, 232
Rodney Start

Page 182
Hazel Richards

Page 200
Christopher Kukura

Horridus looking out over the gallery space at Melbourne Museum.

Triceratops: Fate of the Dinosaurs is now open at Melbourne Museum.